Science in Action

Science in Action

John Lenihan

Illustrated by John B Fleming

The Institute of Physics
Bristol and London

Copyright © 1979 John Lenihan and John B Fleming

Published by The Institute of Physics
Techno House, Redcliffe Way, Bristol BS1 6NX, and
47 Belgrave Square, London SW1X 8QX

British Library Cataloguing in Publication Data

Lenihan, John Mark Anthony
      Science in Action.
      1. Science   2. Technology
      I. Title II. Fleming, John Baxter III. Institute of Physics
      508'1   Q171

ISBN 0-85498-035-0
ISBN 0-85498-041-5 pbk

Filmset in 10/12 point Times New Roman by
The Universities Press (Belfast) Ltd, Belfast, Northern Ireland
Printed and bound in Great Britain by
The Pitman Press, Lower Bristol Road, Bath BA2 3BL

Preface

Science has many images—the instrument of progress, the vehicle for intellectual activity, the agent of environmental calamity, the partner of technology. In today's frantic pursuit of innovation, prosperity and power it is easy to forget that the practice and study of science can be entertaining— even amusing—and that the pleasure can be shared without deep understanding or close involvement. This is not surprising, since science is the study of nature; civilised life—the result of man's interaction with nature—is essentially a cheerful process.

But science seldom stirs the popular imagination in the same way as sport, politics, literature or drama. One reason is that scientists are frightened of the media and unwilling for their ideas to be inflated sensationally or shrunk into trivia. Most scientists turn their backs on the media. Certainly we cannot beat them, but it is interesting occasionally to join them.

Successive editors of *The Glasgow Herald* have generously allowed me the opportunity to learn something of the journalist's craft by writing for their news and leader columns and the feature pages. Many of the essays in this book are based on material originally published in *The Glasgow Herald*. Others are based on articles written for *Books and Bookmen*. I am grateful to the proprietors of both of these journals for their courteous acquiescence in republication here.

Mr Jack Fleming, for many years a regular contributor to *The Glasgow Herald*, has prepared fresh illustrations for the present collection. His witty and elegant drawings are more than decorations, for they show keen insight and are often more successful than the text in conveying significant ideas.

We both appreciate warmly the care and skill applied by Mr Neville Hankins, Miss Valerie Jones and Mrs Terry Poole in the preparation of the final text and in the design of the book.

John Lenihan

Contents

Preface

Why Things Work

History

Science and Society

Earth, Sun and Stars

Sums and Such

The Living World

People

Follies and Conceits

Why Things Work

Freeze, Expansion, and then Friction

In the days when Strathclyde University was still called the Andersonian, the other place in Glasgow was sometimes jocularly known as the Thomsonian, so numerous were the Thomsons among its professors. In the session 1848–9 there were five of them, including William Thomson (the future Lord Kelvin) and his father.

William's brother James had not then succeeded to the Chair of Engineering, but was living in the college pursuing his researches privately. During the winter he developed some subtle speculations on the theory of heat and concluded that an increase in pressure would lower the freezing point of water. The effect did not seem to be very pronounced, since a pressure of about one ton per square inch was needed to change the freezing point by 1 °C, but William, then Professor of Natural Philosophy, made some experiments to verify the calculations.

A stout glass vessel was filled with a mixture of ice and water and a thermometer was inserted. As long as both ice and water are present, the thermometer showed a temperature equal to the freezing point. A plunger at the top of the apparatus was screwed down boldly. Fortunately the glass did not burst, but the temperature of the thermometer distinctly fell. On releasing the excess pressure, the temperature reading went back to its original value.

The lowering of the freezing point by pressure is one of the many curious properties of water and is related to the expansion which occurs on freezing. Because of this expansion ice is less dense than water. Consequently icebergs float, instead of sinking to the bottom, where they would do no harm. On the other hand, the fact that ice floats on water is very convenient, not only for curlers and skaters, but also for the fish, which would not survive the winter if the lochs froze from the bottom upwards.

Winter sports all depend on the slipperiness of snow and ice, a property which has only recently been explained satisfactorily. At about the turn of the century it was suggested that skates move so freely because the pressure of the blades lowers the melting point of the ice and therefore causes some of it to melt, giving a film of water which, of course, freezes again when the skater has moved on.

This explanation is plausible and it survived in the textbooks for half a century, but it is quite wrong. Simple calculation shows that a change in melting point produced by even a heavy skater is not more than a tenth of a degree; yet it is a matter of common knowledge that skating can be done quite readily when the temperature of the air (and the ice) is many degrees below zero.

Some 40 years ago Dr F P Bowden, a Cambridge physicist (portrayed as Francis Getliffe in C P Snow's novels), was skiing in the Alps and spent a few days snowbound in a mountain hut, with ample time for meditation. The skis which had carried him to the hut ran very smoothly at a temperature of −20 °C, though not even an elephant could have produced enough pressure to make ice melt at that temperature.

... snowbound in a mountain hut, with ample time for meditation.

Bowden concluded that the textbooks were wrong and that the smooth motion of the skis over the snow must be attributed to a film of water formed by frictional heat. After that he spent a lot of time in the Alps, combining business with pleasure, in a series of ingenious experiments.

This work led to a large programme of research on general problems of friction and lubrication (now dignified by the name tribophysics), which has led to interesting applications in science and industry.

If the sliding of the skis is really due to a film of water underneath, the friction should increase at very low temperatures, where the snow is not so easily melted. Bowden's experiments showed that this is indeed the case and that the friction on very cold snow (at temperatures of about −60 °C) is almost the same as on dry sand. This effect was known to polar explorers and may have contributed to the loss of Captain Scott's party on their return from the South Pole.

Bowden then began to study ski wax and other materials used to reduce friction. He predicted that PTFE (polytetrafluorethylene, a rather tough, white plastic material) should be even better than wax. This conclusion was verified by further experiments based on measurements of the speed of skis, sometimes loaded with dead weights and sometimes with live enthusiasts.

The results were quite convincing. A ten-stone passenger descended a gentle slope 700 feet long in 83 seconds when mounted on waxed skis of the traditional kind, but took only 54 seconds on skis coated with PTFE. Bowden's discovery has since become widely popular among the skiing fraternity. For some special tasks, such as downhill racing at high speed, the traditional method of waxing may be preferred by experts, but for the skier of average ambition PTFE is a welcome improvement, demonstrating also the soundness of the revolutionary propositions that textbooks can be wrong and that physics is fun.

Unheard Sounds

The Senate of Durham University, sleepily approving syllabuses and degree regulations on a warm afternoon nearly 40 years ago, were roused by an unexpected objection from the Professor of Divinity, Canon A M Ramsey, who later became the Archbishop of Canterbury.

The item to which he took exception was a course of lectures on supersonics. Asked by the baffled scientists to specify the reasons for his concern, he observed in magisterial tones, 'It's a bad word. The correct construction is hyper-acoustics.' The new word did not pass into the university *Calendars* and is not to be found in any dictionary, although supersonics has, by the customary process of escalation, now turned into ultrasonics.

This technology, which deals with sounds too high in pitch to be perceived by the human ear, has many interesting uses in medicine and industry. The first man to study the subject was Francis Galton, an earnest biologist of Victorian times. About a century ago he designed a little brass whistle which gave quite a good output at frequencies well above the audible limit, which is about 15 000 hertz for most people. Galton experimented assiduously in zoos and in city streets, using one of his whistles concealed in a walking-stick and driven by a puff of air from a rubber bulb in the handle. If the animal pricked its ears, it was presumably able to hear the ultrasonic signal. Galton found that cats were best of all, dogs quite good, and insects uninterested.

... cats were best of all ...

After the loss of the *Titanic* in 1912, efforts were made to detect icebergs by sound, by relying on echoes to reveal the presence of large objects which could not be seen in darkness or fog. Little success was achieved in the early experiments using explosions and other loud noises, mainly because the reflected sound was very feeble and could not be easily distinguished from the general background noise, which is quite high in a ship. In 1917, the French physicist Langevin realised the advantage of using sound at a frequency above the audible limit, which would not be subject to interference from the ship's engines and various other disturbances always present in the water. The source of sound that he used was a slice of quartz, made from a handsome crystal which had spent many years decorating a shop window in Paris. A piece of quartz which has been suitably cut shows the property of piezoelectricity. This means that if the crystal is compressed in one direction, an electrical signal will appear in a perpendicular direction. The reverse effect is also found, so that if an alternating voltage is applied to the crystal, it starts to vibrate. If the size of the crystal is adjusted so that its natural frequency of vibration is the same as the frequency of the electrical signal applied to it, the vibrations can be quite large and will generate an intense beam of sound or of ultrasonics. Quartz, and two or three other materials with similar properties, are still used to generate ultrasonics. A crystal of the same kind may be used as a sensitive microphone, giving out an electrical signal when bombarded by an ultrasonic wave.

Many of the uses of ultrasonics depend on two phenomena. The first is that an ultrasonic wave will be reflected whenever it reaches the boundary between one material and another. This effect is the basis of the flaw detector, used both in engineering and in medicine. An ultrasonic signal applied from a quartz-crystal oscillator to the end of a slab of metal will normally pass right through it. If, however, the metal block contains an internal crack, an air bubble, a piece of slag, or any other flaw, some of the ultrasonic signal will be turned back and may be detected by a suitable receiver placed near the transmitter. Alternatively, the transmitting crystal, once it has sent out the initial ultrasonic pulse, may be used as a microphone to pick up any echoes which return.

If an ultrasonic transmitter is applied to one side of the skull a large signal will come back from the other side and a recognisable echo from anatomical structures in the midline will be detected. This test is sometimes applied to accident victims. If the intermediate echo is displaced from its expected position half-way between the two sides, the surgeon may suspect that internal bleeding has pushed the brain out of place. A more sophisticated form of this technique is used to explore the abdomen, particularly during pregnancy.

The second useful property of ultrasonics is cavitation, a process which is still not completely understood, but which is certainly very effective. A sound wave consists of alternate compressions and rarefactions. A liquid can withstand the compressions easily enough, but severe rarefaction will produce an empty space, which will be quickly filled by vapour or dissolved gas emerging from the liquid to form a little bubble. When this bubble collapses a severe shock is generated and may produce unmistakable effects. Engineering components, large and small, may be cleaned very thoroughly by immersion in a liquid bath through which an ultrasonic beam is passed. The shock waves from the cavitation process scour the dirt from the metal surfaces, producing a degree of cleanliness not easily obtained in any other way.

Cavitation, and processes related to it, are exploited by the surgeon when he needs to produce a small region of damage without the risk of injury to surrounding tissues.

Heard melodies (on the authority of John Keats) are sweet, but those unheard are sweeter. The ultrasonic engineer will never be embarrassed by complaints about the noise that he is making, but he can do a lot of good with sounds that no one ever hears.

Water, Water, Everywhere

... nor any drop to drink. The Ancient Mariner's complaint is still heard today, not often in Britain—where the natural wetness provides enough for every tap, with a surplus for the manufacture of more potent drinkables—but quite loudly in other parts of the world.

In ancient times, cities grew up in places where there was a good supply of wholesome water. Town planning today is more subtle and often produces large communities remote from the water that they need. Perhaps it is easier to move the water than to move the people, but the world's thirst is growing fast and shortage of water is now the decisive factor in limiting economic growth in many parts of the world.

... not often in Britain ...

It is unfortunate that the water which covers most of the Earth's surface is undrinkable. One small animal (a desert rat) has been persuaded to drink sea water in the laboratory and has survived without harm, but the human animal is not capable of this achievement. A small amount of sea water can be swallowed with impunity, but anyone who relied on this source would soon die of dehydration. The kidneys cannot deal with such a concentrated solution, and water has to be removed from other parts of the body to dilute the urine to the level at which it can be eliminated.

The purification of sea water is not very difficult as a theoretical exercise, but in practice has been successful to only a limited extent. Desalination needs a certain amount of energy, which cannot be less than the heat of solution of the salts in the sea. When common salt (or almost any other substance) is dissolved in water, a modest amount of heat is given out. No amount of ingenuity will separate the salt from the water again unless the same amount of heat is put back. The heat of solution for sea water is about two-thirds of a calorie per gram. (This is a small calorie, not to be confused with the kind that we eat, which is a thousand times bigger.) If sea water could be purified by this expenditure of energy, the problem would be easily solved. Unfortunately, there is no practical way of doing the job without using a very much larger amount of energy.

The purification of sea water is not very difficult ...

One way of dealing with salt water is to distill it; that is, to boil it and condense the steam, which is then free from contamination and will make very pure water. Sir Richard Hawkins used a crude still to supply drinking water on his voyage to the Americas nearly 400 years ago. When the age of steam arrived, ship's engineers were able to make more complicated devices, and large vessels today usually have distillation equipment aboard.

A great deal of energy (about 540 calories per gram) is required to turn water into steam, and it might seem that the distillation process is too extravagant for large-scale use. However, most of the heat that is lost in vaporisation comes back again when the steam condenses, and it should therefore be possible to design a distillation plant with a rather small energy consumption.

The most successful approach to this problem is demonstrated in the multi-stage flash distillation technique, developed largely through the efforts of Weir Westgarth Limited of East Kilbride, and Professor R S Silver of Glasgow University. Brine is heated and fed into a large vessel where the air pressure has been reduced. Some of the water evaporates and condenses onto pipes near the top of the vessel, from which pure water drops into a collecting funnel.

The latent heat given up by the condensing water is absorbed by the brine flowing through the pipes on its way to the input end of the apparatus. The brine from the first vessel is now passed into a second tank, where the air pressure is still lower and where further evaporation takes place, producing fresh water in the collecting system and contributing to the heating of the incoming sea water supply. This process continues for a large number of stages. The sea water undergoing purification is progressively cooled, but much of the heat taken from it is given up to the fresh intake which consequently needs only a modest amount of additional heating to prepare it for the first evaporation stage.

Equipment of this kind will produce water at an energy expenditure of 50 calories per gram, or even less. The process can be made more attractive (in theory at any rate) if the necessary heat is obtained from a nuclear reactor as a by-product of the generation of electricity, which may also be very welcome in a waterless country.

A still is a noble achievement of technology, producing comfort and inspiration out of mere vegetables, as well as turning sea water into something more drinkable.

Another technique is to squeeze the salt out, instead of boiling it out. If a thin sheet of some suitable material (such as an animal's bladder) is stretched across a vessel, with salt water on one side and fresh water on the other side, some of the fresh water will pass through into the solution. Chemists consider this process in terms of an osmotic pressure, which drives the water into the solution. For this experiment to succeed, the membrane must be semipermeable, allowing pure water to flow through in one direction but preventing the dissolved salts from passing across in the opposite direction.

Is it possible to reverse the process so that by applying pressure to sea water on one side of a suitable membrane, the dissolved salts can be pushed through to the other side? This prospect began to be studied seriously about 20 years ago, when it was discovered that cellulose acetate (a transparent plastic material) is semipermeable to sea water salts.

Early experiments were not very successful because the membrane burst under the applied pressure before a useful flow rate was achieved. By 1960, improved membranes were available and techniques had been found for supporting them so as to withstand the high pressures (half a ton per square inch or more) necessary to give a reasonable flow of fresh water.

Reverse osmosis (as this process is called) should be capable of producing fresh water without a very large expenditure of energy, since the only machinery involved is the pumping gear to bring the sea water to the required pressure. An energy expenditure of six calories per gram has been predicted. Since this energy has to be delivered in the form of electricity (which is itself generated at rather low efficiency), the reverse osmosis technique may not in practice be much more economical than multi-stage flash distillation using waste heat from a nuclear reactor.

In principle it should be possible to separate salt from water just as easily by freezing as by boiling. An ingenious process along these lines has in fact been developed. Salt water is sprayed into a tank where the air pressure has been reduced to a very low level. Some of the water evaporates (because of the low pressure) taking heat from the rest of the spray, which is then partly converted to ice. The ice crystals formed in this way do not contain any dissolved salts and, after washing, can be melted to give pure water.

Some practical success has also been achieved by electrodialysis, in which the unwanted salts are removed by electrolysis. The sodium chloride (common salt) in sea water is present in the form of electrically charged atoms, or ions: some of sodium (with a positive charge) and an equal number of chlorine (with a negative charge).

If a salt solution is put into a vessel containing two electrodes (that is, two rods of metal or carbon) joined to opposite ends of an electrical supply, the positively charged sodium ions will move towards the negative electrode, and the negatively charged chlorine ions will move towards the positive electrode. Suppose now that we put, near the negative electrode, a membrane which is semipermeable to positive ions and, near the positive electrode, a membrane semipermeable to negative ions. The sodium and chlorine ions will disappear through the membranes and the intervening space will before long be occupied by pure water with no salt content. In practice, this process is a little more complicated, since it is necessary (for reasonable economy) to use a large number of membranes, but the principle is the same.

How much is water worth? In Britain we are accustomed to abundant supplies at a cost which may be only a matter of pence per thousand gallons. This excellent arrangement is achieved by a sophisticated nuclear distillation technique. The Sun's nuclear energy is used to evaporate water from the sea for subsequent delivery in purified form as rain. Man-made projects cannot equal the economy of this effort, and the extent to which they will succeeed depends on the cost of the product.

Not so long ago, water was brought to Kuwait by truck and was sold at about two shillings a gallon. By 1970, distillation equipment had been installed to give a supply of nearly 20 million gallons per day at a cost of under £1 per 1000 gallons. This is an acceptable price in Kuwait and in several other Middle Eastern countries, though it is rather high for large-scale industrial or domestic use, and out of the question for agricultural use, except perhaps for specialised crops commanding a very high price.

Nuclear power and water technology will not make the desert bloom like a rose. The trouble is that water, like many of nature's bounties, is very unevenly distributed. So far, however, no one has had very realistic ideas for large-scale improvement of the Creator's efforts in this direction.

Through the Looking Glass

In the bad old days when (according to legend) those responsible for teaching science to little children took their lessons from the textbooks rather than from life, lateral inversion was a familiar topic in the classroom and the examination hall.

Confronted by these two words, the candidate was expected to write half a page explaining how the left-hand side of an object became the right-hand side of the image seen in a mirror. Acceptable answers to this question were evolved over the years and most of those involved did not trouble themselves about the logical absurdities embedded in the proposition.

A bright child might have asked why the image was not upside down as well, since it is reasonable to expect that the laws of optics will be the same in a horizontal direction as in a vertical direction. A few of the bolder textbooks came close to admitting that lateral inversion was a fraud, but generations of pupils were brought up to believe in it.

Through the looking glass.

To dispel this trickery we need to consider what we mean by an image in a mirror. When light strikes a solid surface, such as a brick wall, most of it is absorbed (and eventually turned into heat), and the rest is reflected, though not in any particular direction. A highly polished surface, such as a mirror provides, absorbs very little of the light and reflects the rest. The reflection is more orderly than that from a brick wall.

Imagine a small source of light, such as a torch bulb, in front of a mirror. Draw lines from the source to the mirror, representing two of the many directions (or rays) in which the light is emitted; each of these rays will be reflected. The important property of a mirror is that the reflection is symmetrical; in other words, the reflected ray is inclined to the mirror at the same angle as the incident ray. For this reason, the two reflected rays appear (to an eye situated in front of the mirror) to be coming from a position behind the mirror, where the image of the torch bulb is seen. It is not difficult for those versed in Euclid's geometry to show that the image is as far behind the mirror as the object is in front of it.

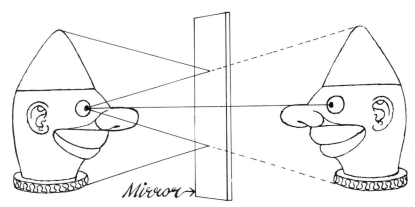

The image is as far behind the mirror as the object is in front of it.

Suppose now that we take a more complicated object. Write a word on a piece of paper, hold it up to the mirror and see what sort of image is produced. The image seems to be the wrong way round and is not very easy to read. This alteration is not really the fault of the mirror. Remember that when you write the word on the paper, you have to turn it through an angle of 180° in order to present it to the mirror. That is why the image appears to be reversed. Write the word on a piece of transparent paper and hold it to the mirror in the normal way; you will then see that each letter of the image corresponds in position and shape to the appropriate letter of the object.

People who have tried to knot a bow tie in front of a mirror may still complain that the task is difficult because the left and right hands of the image appear to be interchanged with the originals. This is not really what happens.

We have already seen that the image is as far behind the mirror as the object is in front of it. Stand in front of a tall mirror and consider how the image is formed. The nose, chin and toecaps (or in some people the watch chain) are nearest to the mirror and their images will therefore also be nearest to it, though on the opposite side. The ears and arms are further back and their images will be correspondingly further from the mirror. If you raise your left hand, the image hand straight in front of it will also rise. If you have a ring on your right hand, the image of it will be seen at a point straight in front of the ring.

Why do we think that the left and right hands are interchanged? Suppose that the mirror is on a north-facing wall. If you stand in front of it, a line drawn from the back of your head through the tip of your nose runs from north to south. But a line from the back of the image head through the tip of the image nose runs from south to north.

What the reflection really does is reverse the relative positions of points lying on a line perpendicular to the mirror. Since it is a matter of common experience that the nose comes in front of the face, our eyes and brains react as though the image was a person facing in the opposite direction to the original object. This illusion can be appreciated by remembering that when two people shake hands, each crosses his right hand in front of his body. But there is no need to do this if you want to greet your image in the mirror; just hold up your right hand, press it flat against the mirror and you will see that the image hand meets it at every point.

Man Against the Floorboards

Why do we not fall through the floor? The study of this serious question leads into many fascinating areas of science, engineering and history†.

The story begins in 1687, with Isaac Newton's assertion that action and reaction are equal and opposite. A 12-stone man standing on the floor presses the boards with a force of 168 pounds. The reason why he does not fall through to the cellar is that the floor presses upwards on the soles of his feet with exactly the same force. Why does the floor act in this aggressive way? A lump of wood in its normal state shows no tendency to push people about or to exert any pressures that an observer could notice.

The reaction in Newton's famous law comes about because every material is elastic in the technical sense; in other words, not utterly rigid. A load applied to any material causes a deformation, which is resisted by its natural elasticity. The restoring force, by which the loaded object seeks to return to its original state, depends on the extent of the deformation. The floorboards yield until the restoring force is just equal to the weight of the load.

Permanent structures have to be made from materials which can generate substantial restoring forces at very small deflections. Lean against a brick wall and the compression will be no more than a few millionths of an inch. On the other hand, the suspension cables of the Forth road bridge, for example, are permanently stretched to ten feet more than their normal length (about two miles) in order to support the load provided by the roadway and the traffic.

Structures can be designed so that the materials from which they are made are in compression (like the floorboards), or in tension (like the cables of the bridge). Among the common building materials, stone is strong in compression but weak in tension. For this reason, medieval architects were careful to see that the blocks of stone in their cathedrals and palaces were squeezed but never stretched. A stone bridge arch is a good example of this principle at work. The wedge-shaped blocks of stone forming the archway itself are compressed by their neighbours on either side, and by the roadway, rubble, and earth-filling from above. To maintain adequate compression, Gothic cathedrals often had to be propped up by buttresses; sometimes the squeezing was too vigorous and the internal walls had to be held apart by struts or inverted arches to prevent collapse. Timber is quite strong in tension, but was not fully exploited in earlier times because of the difficulty in making joints of comparable strength.

The Greeks did not care for arches, but preferred beams and columns. The Parthenon and other Doric temples are essentially wooden

† J E Gordon 1968 *The New Science of Strong Materials* (Harmondsworth: Penguin).

buildings copied in marble. Because of the limited tensile strength of this material, the gap bridged by a single slab was seldom more than eight feet wide. The gateway to the Acropolis had to be considerably wider, with a span of up to 20 feet; the architect tackled this problem by cementing large iron rods into grooves in the marble. Later attempts at reinforcement often failed because water seeped in and rusted the metal.

The Greeks ... preferred beams and columns.

In about 1850, it was discovered that iron buried in concrete does not rust to a dangerous extent. Unfortunately, concrete is a weak material when stretched or bent, and is liable to serious cracking if loaded in the wrong way. The solution to this problem is to keep the reinforcing rods stretched while the concrete is setting. In this way the concrete is put into a permanent state of compression. Pre-stressed concrete has been available since 1890, but it has only recently been exploited on a large scale.

Wood is an excellent material for the amateur constructor. The cell walls in its fibres collapse rather easily under compression. It is for this reason that nails and screws can be used so successfully.

On a larger scale, the designer cannot always predict with certainty every squeeze and stretch that will be incurred throughout a structure. Iron, steel and reinforced concrete, being strong in compression as well as in tension, are therefore very useful materials. Unfortunately, however, their resistance to fracture or collapse does not approach the limits suggested by simple and reliable calculation. Steel is unusual in that its measured strength is sometimes as much as 10% of the theoretical value; for many other common materials the corresponding figure is less than 1%. A clue to this deficiency can be found in the observation that thin wires are usually stronger (weight for weight) than solid lumps of the same material. If a glass rod is stretched, it will break when the stress reaches about 25 000 pounds per square inch, but a very thin glass fibre can be stretched to nearly 100 times this stress. A A Griffiths, a pioneer in this work, pointed out nearly 60 years ago that the great strength of thin fibres is not at all mysterious, since in the ultimate limit a single chain of atoms must have either the strength predicted by calculation or none at all. The real problem is to explain why thicker samples are so weak.

A thread of glass, less than a thousandth of an inch thick, is as strong as steel, by any test that an engineer chooses to apply. Why is it so strong? Not because it is thin, and not because it is glass, but mainly because it is smooth.

The inherent weakness of most materials is due to the presence (and rapid propagation under stress) of cracks. A brick is full of small voids, and a slab of cast iron, even if it has been carefully made, contains many veins of graphite which act in the same way. In other materials, the cracks may be of only atomic dimensions.

Any material will be very strong if it can be kept free of cracks. Glass is easily scratched. Microscopic inspection always shows a multitude of cracks, even when none is visible to the naked eye. When in contact with an obstacle, thin fibres are more likely to bend than to stay put and be scratched. A thick glass rod would be as strong as a thin fibre if its surface could be kept as smooth.

The mechanical properties of a brittle material such as glass are quite satisfactory so long as no great stress is applied. A wineglass or a mirror will last indefinitely in the absence of serious disturbance; but drop it on the floor and it will be shattered, as the tiny cracks always present on the surface open up and spread with explosive violence.

Brittle materials are still useful if they can be kept in compression

... so long as no great stress is applied.

rather than in tension. The toughened glass windscreen, commonly found in modern cars, provides an example of this technique. A sheet of hot glass is cooled by air jets. Since glass is not a very good conductor, the outside layers cool while the inside remains hot. The outside of the glass contracts as it cools so that the inside, being at a higher temperature, occupies a greater volume. In this way the outer surface of the glass is put into compression, while the inside mass solidifies to occupy a volume greater than normal, with the result that a state of tension is established. Toughened glass will stand up to minor insults, but if a crack does penetrate the outer layer the whole structure will break into thousands of tiny crystals.

Strain patterns in glass are revealed by polarised light, and can often be seen without any specialised equipment, since the light from a blue sky is partly polarised, particularly when the sun is low. Alternations of light and shade corresponding to the positions of the air jets used in cooling the glass can sometimes be seen when a windscreen is viewed at the correct angle.

Many of the properties of glass are linked with its peculiar physical structure. Metals are crystalline solids, whose atoms are arranged in regular rows and columns. This much is known from experiments in x-ray crystallography.

A good echo can sometimes be obtained from an open fence if the pitch of the sound is chosen so that the time interval between reflections from successive uprights is correct for reinforcement. In a similar way the scattering of an x-ray beam by a solid specimen placed in the path gives information about the way in which the atoms are arranged.

X-ray studies of glass show that the atoms are arranged in a disorderly way, characteristic of a liquid and not, as might be expected, in the regular pattern of a crystalline solid. Glass is really a liquid with very high viscosity. Most liquids crystallise when cooled to just a few degrees below freezing point. Glass is so sluggish (because of its unusually high viscosity) that the molecules cannot rearrange themselves in the regular pattern needed for crystallisation. Consequently, as molten glass cools, it retains all the characteristics of the liquid state—except, of course, that it looks and feels like a solid.

When molten sugar is cooled fairly quickly, it solidifies to a kind of glass (familiarly known as toffee) which is very useful for experimental purposes. Stress it cautiously and it can be bent double; but give it a sharp tap with a hammer, or on the edge of a table, and it snaps. Many tough materials, including pitch and some plastics, behave in the same way. Slowly applied stress is relieved as the material flows away from the point of pressure; but the crack produced by a sharp blow grows too quickly to be dispersed by plastic flow.

Glass does sometimes crystallise, especially if it is very old; the atoms have then had plenty of time to rearrange themselves in a disciplined way. Crystallised (or devitrified) glass is weak, tending to break along the crystal boundaries; devitrified toffee is tablet—or fudge, as Professor Gordon calls it.

Hot and Cold

Talleyrand, when confronted by someone he did not know well enough to engage in meaningful conversation, would open the exchange by asking: 'How's the old trouble?' This invitation invariably released reminiscences and complaints, providing an adequate description of the speaker and a sound basis for further discussion.

Lord Snow, physicist and novelist, in his excursions between the two cultures, devised different methods. Turning graciously to his neighbour at the dinner table, he would ask: 'What do you know about the second law of thermodynamics?' More recently, he is said to have abandoned this enquiry (perhaps because young ladies have become better informed on scientific matters, but more probably because hostesses have become more cunning), in favour of a question about nucleic acids.

'What do you know about the second law of thermodynamics?'

In less exalted circles, thermodynamics remains a serviceable topic for enquiry. It turned up the other day when a group of philosophers was relaxing over coffee during the interval between the mid-day carbohydrate and the afternoon committee. Someone asked for an explanation of the absolute zero of temperature. A cautious colleague asked whether anyone could define temperature for a start. 'Degree of heat,' 'Kinetic energy of molecules' and 'Just an abstraction' were among the suggestions quickly formulated; all of which are quite reasonable, for the concept of temperature has its roots in primitive technology and its branches in the airy realm of thermodynamics.

The abstractions that form the raw material for the progress of science are always derived from observation. The concept of hotness (which is not the same as heat) was familiar in ancient times. Aristotle asserted, 2300 years ago, that all matter possessed, in varying proportions, four distinct qualities described by the adjectives hot, cold, moist and dry. Alchemists, physicians and philosophers found this classification useful in a great many ways, even though they had no system of units or measurements to give numerical expression to their speculations.

Galileo is sometimes credited with the invention of the thermometer, but the instrument which he devised, in about the year 1592, would hardly be recognised today. As his friend Benedetto Castelli described it many years later:

> He took a glass about the size of a small hen's egg, fitted to a tube the width of a straw, and about two spans long: he heated the glass bulb in his hands and turned the glass upside down so that the tube dipped in water held in another vessel; as soon as the ball cooled down, the water rose in the tube. This instrument he used to investigate degrees of heat and cold.

Galileo's thermoscope depended on the observation that air expands when it is hot and contracts when it is cool.

The water did not act like the mercury in a modern thermometer, but merely served to indicate the expansion or contraction of the air in the bulb. The preliminary heating was a trick to bring the water to a level where it could conveniently be seen.

The first man to make the thermometer into an instrument of practical value was probably Sanctorius, Professor of the Theory of Medicine at Padua from 1612 to 1624. He was the first physician to take a patient's temperature and to realise that a thermometer could give information useful in the diagnosis and the management of disease.

The next decisive improvement in design was made by Ferdinand II, Grand Duke of Tuscany, in about the year 1641. His thermometer, the first to be independent of changes in atmospheric pressure, was based on the expansion of a liquid, rather than of air. A glass bulb, filled with coloured alcohol, led into a tall, narrow tube sealed at the top.

The early thermometers sometimes bore scales with numbers or other marks to indicate degrees of heat and cold. The ingenious Burgomaster of Magdeburg, Otto von Guericke, built a magnificent specimen, 20 feet high, in which the expansion of alcohol moved a float connected to the figure of an angel pointing, as the weather dictated, to divisions graduated from 'great heat' to 'great cold.'

One of Ferdinand's thermometers reached Robert Boyle, who observed: 'We are greatly at a loss for a standard whereby to measure cold. The common instruments show us no more than the relative coldness of the air.' Boyle's attempts to define six points on a thermometer were not very successful, but a significant advance was made in 1694 by Carlo Renaldini, a mathematician of Padua, who suggested that the melting point of ice and the boiling point of water should be used as standard temperatures, the space between them on the thermometer stem being divided into twelve parts. This was a sound idea, but it was not taken up for many years.

Daniel Fahrenheit made the first reliable mercury thermometer early in the eighteenth century. On his scale, to which the British are still attached, the melting point of ice was placed at 32 degrees and the normal temperature of the body at 96 degrees. The upper fixed point was afterwards identified with the boiling point of water at 212 degrees.

People devoted to British weather often believe that the centigrade scale was devised by riotous Frenchmen, along with kilograms, centimetres, and other unwelcome manifestations of Gallic perversity. The notion of a scale of temperature with 100 degrees between the melting point of ice and the boiling point of water was actually proposed by Anders Celsius, a Swedish astronomer, in 1742.

Although the thermometer is an apparently simple instrument, its proper understanding requires an excursion into the subtleties of thermodynamics.

Beethoven (wrote the schoolboy) composed four symphonies: the Eroica, the Fifth, the Pastoral, and the Ninth. A similar confusion surrounds the laws of thermodynamics, which govern all the transformations of energy which occur in the body, in the fireplace, and throughout the universe.

The first law is very familiar. The second law is known to many people (including Lord Snow), and a company of earnest students will, if provoked, write half a page about the third law; but even the most costly textbook goes no further. Before revealing where the fourth law is hidden we should look at the other three, since they point the way to the absolute zero which the lunchtime philosophers were discussing when we left them a page or two back.

Each of the laws can be stated in many different ways, some more rigorously than others. A simple version of the first law says that energy can neither be created nor destroyed. There is some justification in the criticism that this law is maintained only because scientists invent a new kind of energy to deal with any apparent violation. Strike a match, and a great deal of energy is produced in the form of heat and light. The physicist explains this away by insisting that the head and stalk of the match are stuffed with chemical energy, waiting to be transformed into more obvious varieties. Needless to add, the chemical energy is defined in such a way as to make the calculations come out correctly, indicating the same total amount of energy before and after ignition.

In some processes, such as nuclear fission, it is necessary to include on each side of the balance sheet a term representing the mass of the materials involved. This refinement was introduced in 1905 by Einstein, embodied in the celebrated rule: $E = mc^2$, indicating the amount of energy E which can be obtained from a mass, m; where c is the speed of light.

The second law of thermodynamics draws our attention to some special considerations about heat, which do not apply to other forms of energy. It is a matter of common observation that if two objects at different temperatures are placed in contact, heat will flow from the hot body to the cold body until they both have the same temperature. This process is the basis of an experimental exercise, known to generations of students as the method of mixtures. It was not realised until late in the day that the eventual equality of temperature between two bodies in contact was fundamental to the three well established laws of thermodynamics. The long-known but newly elevated principle was therefore called the zeroth law of thermodynamics.

Although the concept of temperature is quite familiar, a satisfactory definition of the word is not so easy. The scientist is tempted to follow Humpty Dumpty: 'When I use a word, it means just what I choose it to mean.' The temperature of a flame might be defined in terms of its colour; the temperature of a patient by the length of the mercury thread in the physician's thermometer; and the temperature of a coil of wire in terms of its electrical resistance. The scientist, however, looks for definitions which will not be linked so obviously with the properties of particular substances or materials. One way is to define the temperature of a body in terms of the average kinetic energy of its molecules. When the kettle is switched on the electrical energy from the supply is converted into heat, and is then shared out as additional kinetic energy among the water molecules, which consequently rush around at increasing speeds, until eventually they acquire enough energy to burst out of the liquid state altogether and escape as vapour.

When a hot and a cold object are placed in contact the molecules exchange energy in the collisions which occur at the points of contact. After a sufficient number of collisions the molecular energies will be evened out, and the temperatures of the two objects will approach the same value.

The zeroth law of thermodynamics is of great practical importance. Strictly speaking, a thermometer measures its own temperature; it is reassuring to know that if we leave it for long enough to retain thermal equilibrium it will record accurately enough the temperature of the air, the body, or any other object in contact with it.

For scientific purposes the gas thermometer is one of the best temperature-measuring devices. If some gas is put into a container and heated, its pressure increases; the pressure is merely the battering of the gas molecules on the walls of the container, and this obviously becomes greater as the molecules gain higher speeds through being heated. Conversely, if the gas is cooled, the pressure which it exerts will become less. Students in the physics laboratory often measure the pressure of a gas at various temperatures. By drawing a graph and producing it backwards they can predict the temperature at which the gas would exert no pressure at all. For all the common gases this temperature is about $-273 \cdot 2\,°C$, which is known as absolute zero. The approach shown by the student's graph is not, of course, entirely realistic, since the gas will change to a liquid and then to a solid before reaching its lowest possible temperature. Absolute zero can, however, be defined in a variety of ways, and it always occurs at the same place on the centigrade scale.

How can we reach or approach absolute zero? The second law of thermodynamics tells us that heat will flow freely from a hot body to a cold body, but it will not flow unassisted in the reverse direction.

High-speed gas will boil a kettle of water, but the water will never turn spontaneously into a block of ice so that the flame may burn a little hotter. Even though heat will not flow spontaneously from an object at a low temperature to one at a high temperature, it can be encouraged to do so if an additional source of energy is available. This is the underlying principle of the refrigerator.

The approach to absolute zero is made by refrigerators of increasing subtlety. The third law of thermodynamics—based, like the other three, on common experience (or at any rate on the experience reported by reliable scientists)—tells us that we can approach absolute zero but can never reach it, however great the experimental ingenuity applied to the problem.

Thermodynamics arose out of attempts to explain the steam engine, but did not wither when that important problem had been solved. On the contrary, the subject became more subtle and sophisticated, spreading into every branch of science and engineering, and it is still quite a lively area for research. But the man who discovers the fifth law of thermodynamics will have great difficulty in finding a suitable name for it.

A Shocking Story

Aptly situated in a quiet street, near the American Embassy in London, stands a modern hotel offering many comforting examples of technology: warm air, fire-proof doors, and wall-to-wall carpets owing nothing to sheep. An unexpected property of this agreeable environment is the electric shock which the guest receives whenever his hand or key touches the lock.

This hazard is not very familiar in Britain, a country protected by wool, cotton and general dampness, but its origins go back to the very beginning of electrical science.

... the electric shock which the guest receives whenever his hand or key touches the lock.

The shocks which make the hotel guests jump are produced by static electricity. In the days before factories and power stations, the production of electricity was quite an adventurous business, giving great scope for the ingenuity of the natural philosophers. The earliest students of electrical science worked in Syria more than 3000 years ago. The women spun their yarn in traditional fashion, with a distaff in the left hand to hold the raw wool, and a spindle in the right hand, twisting the loose fibres into a strong thread as they fell. The spindle was often made of amber and the spinsters, noticing how often it attracted fluff and dust, gave it the name *Harpaga* (the clutcher), a word carried into English language by the harpies of later times.

After this intelligent observation, progress was slow. Materials which displayed the curious property noted by the spinning women became known as electrics, from the Greek word for amber, but there was much confusion between electricity and magnetism; St Augustine was one of the few scholars who appreciated the difference.

The first machine for the continuous production of electricity was made about 300 years ago by Otto von Guericke, the Burgomaster of Magdeburg. His machine consisted simply of a ball of sulphur 'about the size of the head of an infant' which became electrified when rubbed with the hand. The electric globe, sometimes made in more elaborate forms and rotated mechanically, became a popular instrument of instruction and entertainment.

Stephen Gray was a pensioner of Charterhouse. Not content with munching and meditation, he spent his time in electrical investigations, with the help of boys from the nearby Grey Friars School. In the spring of 1730, he made a capital experiment by suspending a boy with two silk ropes, applying the electrical influence to his feet, and drawing sparks from his face. This observation was exploited very significantly a few years later by Charles Du Fay, a French diplomat and amateur of science.

Electricity, Du Fay said, is a fluid. It can flow freely through metals but is restrained by other substances, which he called electrics, but which we should now describe as insulators. If a charge is produced on an insulator, it stays there and constitutes static electricity. A charge can be produced quite easily on a conductor, such as metal, but it usually leaks away very quickly. If, however, a metal object is carefully insulated (from its surroundings) it will hold an electric charge just as well as a piece of amber or a ball of sulphur.

The electric fluid, Du Fay continued, can be of two kinds. Resinous electricity is produced, for example, by rubbing wax, and vitreous electricity is produced by friction on glass. The electric charge can be transferred from one object to another merely by contact. If two objects are both charged with the same kind of electric fluid they will repel each other; but if one is charged with vitreous electricity and one with resinous electricity, a force of attraction will be observed.

In 1747 Benjamin Franklin asserted the surprisingly modern view that electricity comes in only one variety. Vitreous electrification, he said, is merely a surplus (alternatively described as a positive charge) and resinous electrification is a deficiency, equally well classified as a negative charge.

A century and a half later, the one-fluid theory of electricity was conclusively established by experiment, but the basic unit turned out to be the electron, a particle which, according to Franklin's classification, must be allocated a negative charge.

Shortly before J J Thomson's discovery of the electron in 1897, H A Rowland of Baltimore made another decisive experiment. He put an electric charge on to an ebonite disc, spun it rapidly, and showed that a nearby compass needle was deflected. In this way he demonstrated what had long been suspected, that an electric current is merely a movement of electric charge.

By the end of the nineteenth century, electricity was no longer the mystery that had intrigued the philosophers and craftsmen through the ages; but the old-fashioned static electrification was to become even more important during the twentieth century. The scene was set by an amateur:

Maryland, Anno 1653

There happened about the month of November to one Mrs Susanna Sewall, wife of Major Nicholas Sewall, of the Province aforesaid, a strange flashing of sparks (seem'd to be of fire) in all the wearing apparel she put on, and so continued to Candlemass: and in the company of several, viz., Captain John Harris, Mr Edward Braines, Captain Edward Poneson, etc., the said Susanna did send several of her wearing apparel, and when they were shaken it would fly out in sparks and make a noise much like unto bay leaves when flung into the fire ...

... a strange flashing of sparks ...

Mrs Sewall was an early victim of static electricity, though she seems to have used the ordeal to good purpose in entertaining her gallant friends. At a subsequent session she exchanged petticoats with her sister, Mrs Digges, and gave a further display of luminosity, constituting the first electrical experiment performed in the New World.

Using the right materials, which are now readily available, it is not difficult to produce sparks when casting undergarments or even when combing the hair. The effects which so greatly surprised the early settlers in Maryland involve two processes. The first is that electric charges are nearly always produced when two solids are placed in contact and then separated. The electrification is often enhanced by rubbing the materials together before parting them, but charges can be obtained without any friction at all.

Many substances, including metals, common building materials, and the tissues of the body, allow fairly free movement of electric currents. Consequently, charges which may be produced by friction or contact usually leak away to earth in a short time.

Charges produced on insulating materials stay put and can accumulate almost indefinitely. The limit is reached when the electric field around the charge becomes great enough to break down the insulation of the air separating the charged object from the nearest earthed conductor. The charge then disappears, often with a visible spark. If the charge is on a person's body he may experience a perceptible shock, corresponding to the flow of current in the spark.

As we walk, electric charges are produced on the floor and on the soles of our shoes. The charge on a shoe is usually dissipated as the foot touches the floor and reappears as the shoe becomes airborne. If the floor is covered with a particularly good insulator, such as a carpet of nylon or some other man-made fibre, the charge does not escape but spreads over the pedestrian's skin and clothing. As soon as he touches an earthed object, such as a door knob or a light switch, the charge takes the easy way to earth, probably giving him a slight shock in the passing.

In many circumstances this effect is merely annoying, but it can sometimes also be dangerous. The spark produced by the disappearance of static charge is enough to cause an explosion if it occurs in the presence of certain anaesthetic gases, such as ether.

Explosive anaesthetics are not often used in modern surgery, but operating theatres always include many precautions against the build-up of static charges. Plastic materials and man-made fibres are avoided. The rubber used in wheels, tubes and other appliances is of a specially prepared kind with reasonably good electrical conductivity, so that static charges cannot accumulate. If the floor is not of terrazzo, its electrical properties have to be checked in advance.

Static charge is a nuisance in the manufacture of plastic sheeting, paper and man-made fibres. The friction which is inevitable in the production of these materials generates considerable charges which may cause sheets or threads to fly apart in an uncontrollable way. As the Syrian women discovered more than 3000 years ago, electrical objects attract dust which may spoil the appearance of a woven fabric. In extreme instances, the sparking associated with static electrification may cause fires or explosions in factories.

The way to deal with this problem is to ionise the surrounding air, that is, to fill it with electrons and positively charged atoms. If the static electrification is negative in sign, the charged surface will attract positive ions from the air and return to a neutral electrical status. A positive charge of static will be dissipated in a similar way by electrons attracted from the surrounding atmosphere. As long as the supply of ions and electrons is maintained, potentially dangerous accumulations of static charge will be quickly dispersed.

A simple way to produce ionisation is to place a suitably shaped radioactive source near to the electrified object. The radiations emitted continually from the source will cause ionisation in the air and (if a long-lived isotope is chosen) will work for years without attention.

Static electrification is not always dangerous. The efficacy of paint spraying is greatly improved by giving the droplets a substantial electrical charge, so that their mutual repulsion spreads them more evenly over the surface; the same principle is used in crop spraying.

An intriguing application of this technique, which was pursued for several years in Scotland, concerned the manufacture of kippers. The traditional process is not very efficient, mainly because the smoke is not uniformly distributed over the fish. The new approach operated by blowing the smoke through a grid of wires charged to a high voltage. The smoke particles, all electrified, were then admitted to the kippering chamber, where they spread uniformly in every direction under the action of the electrical repulsion.

In the technical sense, this method was a great improvement on the old-fashioned process which—whatever the romantic legends may say—was often haphazard. Unfortunately, it was difficult to demonstrate the superiority of the scientifically prepared kipper, either to the trade or to impartial tasting panels. In some countries, however, electrified smoke is used to improve hams, turkeys, and even sardines.

Two Wheels are Best

As the motor car dies of thirst the bicycle is returning to the roads. Although its main attraction is the ability to run on steak and claret, fish and chips, or any other fuel that the rider enjoys, the bicycle is also a masterpiece of technology. Some of its many remarkable features have only recently been recognised and some are still not fully understood.

The wheel was probably invented 5000 years ago, but was not developed into a convenient personal transport system until well into the nineteenth century, even though the necessary materials and techniques had been available for some time. The undoubted inventor of the bicycle was Kirkpatrick Macmillan, the Dumfries blacksmith whose treadle-driven two-wheeler took to the road in 1839.

This machine was not a great commercial success, but the velocipede (made in France in 1863), became very popular. The pedals drove the front wheels directly. Consequently, a full turn of the pedals moved the rider a distance equal to the circumference of the wheel; this meant that the cyclist had to pedal furiously to reach a reasonable speed, for the machine was, in modern terms, low-geared.

The first solution—not a very clever one—to this problem was to increase the size of the front wheel, thus producing the penny-farthing design. The modern bicycle dates from the introduction of the chain drive (in 1879), allowing any desired gear ratio with wheels of convenient size. The Rover bicycle, made in Coventry in 1885, had chain-driven ball-bearing hubs and a tubular steel frame; in fact, all the essentials of the modern machine, except for pneumatic tyres which were added a few years later.

The bicycle is by far the most efficient means of transportation. As a system for moving with minimum expenditure of energy, the combination of man and bicycle performs better than any living creature or machine.

... by far the most efficient means of transportation.

The easiest way to compare the efficiencies of different systems is to calculate the energy used in moving one gram through a distance of a kilometre. Walking at a normal pace a man uses about 3 joules per gram weight per kilometre travelled. This is very much better than a rabbit or a helicopter, about the same as for a motor car, but not quite as good as a jet aircraft.

When the man is mounted on a bicycle his energy consumption is reduced to 0·6 joules per gram weight per kilogram moved—a figure not approached by any travelling animal or machine. By taking to two wheels a man can (allowing for increased wind resistance) travel two or three times as fast for the same expenditure of energy.

The reasons for this spectacular improvement are not hard to see. Walking is a good way of covering rough ground, but it is a rather wasteful process. Even when standing still we use a modest amount of energy in keeping the leg muscles tense to support the rest of the body. In walking, a good deal of energy is wasted in raising and lowering the body, as well as in friction when the feet touch the ground.

The cyclist is better organised. As he is sitting he does not use much energy in maintaining the posture of the body. His legs and feet move at a fairly uniform rate, avoiding wasteful acceleration and deceleration. The frictional losses are greatly reduced by the substitution of wheels for feet.

To be fair, the energy advantage is not quite as big as it seems. It might be thought that human muscle power is a bonus quite independent of the energy crisis. This is not so, for the production of food uses an ever-increasing amount of fuel; for example, in driving farm machinery and in producing chemical fertilisers and pesticides. But the extra fuel consumption required for cycling is much less than would be used in any other form of transport. If people must move about, cycling certainly makes the minimum demand on global fuel resources.

The bicycle, like the steam engine, is a good example of the way in which technology can succeed without the benefit of scientific guidance or understanding. Indeed, if the bicycle had been designed by scientists (instead of blacksmiths) it would probably never have worked. Even now the remarkable stability of the bicycle is difficult to explain.

Dr David Jones, an English chemist, studied the problem during the 1960s. A riderless bicycle, he observed, will fall over within a second or two if left alone; but if pushed and released it will stay upright for as long as 20 seconds, moving in a gentle curve before collapsing. It is a matter of common experience that a ridden bicycle is very stable, especially at high speeds. Dr Jones asked why, and made some interesting experiments in his search for an answer.

One obvious solution is that the front wheel acts as a gyroscope, with the inherent stability shown by the hoop, which was a familiar toy before children became affluent. Is the bicycle a hoop with the rear wheel merely trailing behind?

Dr Jones made a test by fixing an additional front wheel on the same axis, but slightly smaller so that it was clear of the ground. If he spun the extra wheel in the same direction as the road wheel, the riderless bicycle was more stable than ever; but if he cancelled the gyroscopic effect by spinning the loose wheel in the opposite direction the machine promptly collapsed. However, when he rode the modified bicycle the extra wheel had no effect whichever way it was spun. So a bicycle is essentially a hoop, stabilised by gyroscopic forces, but only while riderless.

What keeps a bicycle upright when ridden in the normal way? One of Dr Jones's colleagues suggested that the stability is provided by the width of the tyres—in other words, that the bicycle is a thin steam roller—but this idea is not very attractive on further study. Dr Jones tested various other theories by trying to build unstable bicycles, but they all turned out to be quite rideable.

Next he programmed a computer to design a bicycle, and eventually found how to make an unstable machine by extending the fork and moving the front wheel four inches ahead of its normal position. This device was very difficult to control and completely unstable when riderless. The computer also showed that the Lawson Safety Bicycle of 1879 was well named, for it was more stable than any subsequent commercial model.

We still do not know why the bicycle works so well; but perhaps some enterprising university is already making plans for a Chair (or Saddle) of Kuklosophy to find the answer.

Science and the Bagpipe

The Bagpipe is practically an obsolete instrument and only found now in the hands of beggars and the indigenous population of England, Scotland and Ireland.

Dr Hugo Riemann was a professor in the University of Leipzig and a musicologist of great renown, even if he was not very well informed about the bagpipe when he wrote this entry in his *Dictionary of Music* in 1895.

Riemann was not alone in his ignorance and lack of perception. Some of his contemporaries heard the bagpipe well enough, but regretted the distinctive character of its music and suggested that a great improvement would be achieved by conformity with the diatonic scale favoured by less imaginative composers and executants.

Dr W H Stone explained in 1879 that the piper's grace notes introduced to conceal the deficiencies of the bagpipe scale

> ... are termed warblers, very appropriately, after the birds, who until trained and civilised, sometimes by the splitting of their tongues, entirely disregard the diatonic scale.

In 1940, Dr G E Allan of Glasgow reported the results of a long investigation of the bagpipe scale, and concluded with the suggestion that the chanter should be re-designed to produce the notes of the diatonic scale for

> ... if the untunefulness of the scale in its present form were removed, so also would the irritation which it causes to musical people, whereas non-musical listeners would not notice the change.

'... are termed warblers, very appropriately, after the birds.'

This proposal owes something to the traditional belief that foreigners speak gibberish because they do not know any better. People who object to the bagpipe because it is untuneful will generally have the same complaint about Arabic, Indian or Chinese music, and for the same reason. As the proverb might have said, familiarity breeds content.

The *ceol mor*, or big music, has survived centuries of abuse, indifference and incredulity because in Scotland the Highland bagpipe, and those who play it, have resisted all pressures for change.

Yet the bagpipe is not peculiarly Scottish, and was probably known in other lands long before it appeared in Scotland. The earliest evidence, from a Hittite carving of the thirteenth century BC, is dubious, but there is no doubt that the brave sound was familiar to the Romans, whose soldiers brought the pipe to England—and perhaps to Scotland. According to Suetonius, Nero played the bagpipe.

Varieties of bagpipe appeared in many parts of Europe during and after the heyday of Rome, but the cult declined after the end of the Middle Ages—except in the Highlands of Scotland.

The survival of the Highland bagpipe, now played enthusiastically in France, North America, India and Pakistan (where one of the Government's first actions after independence was to commission an appropriate arrangement of the national anthem), is attributable to three causes.

Firstly, the Middle Ages lasted longer in Scotland. Houses were mere shelters and most of the activities of life were pursued in the open air, where no musical instrument could compete with the pipe. Secondly, the Highlanders resisted the temptation to abandon their heritage by making smaller and quieter pipes, suitable for indoor use. Thirdly, the persistence

... the Romans, whose soldiers brought the pipe to England—and perhaps to Scotland.

and eventual popularity of the Highland bagpipe are owed to the distinctive character of the *piobaireachd*†, its classical music.

Bagpipe music comes in several kinds. The marches, strathspeys and reels so often played by bands are classified as *ceol beag*, or little music. But the distillation of Scotland's life and history, the authentic record of the nation's pride, is the *ceol mor* (or *piobaireachd*), which is as far removed from other forms of pipe music as a sonata is from a pop song. *Piobaireachd* is impossible to define in a few words. To dismiss it as a theme with variations is no more perceptive than to describe a Shakespeare play as a collection of speeches divided into five acts.

To the inexperienced ear—that is, to most of the world—bagpipe music sounds strange. There are several reasons for this. The piper has no control of the loudness of the instrument. He can produce only one note at a time and, once started, cannot stop until the end of the composition that he is playing. The scale is also unique, although its characteristics were defined in scientific terms only in 1953.

The important features of a musical scale are the pitch of the notes and the intervals between them. Standard orchestral pitch is based on $A = 440$ hertz (or cycles per second, as it was formerly and more aptly described). In the bagpipe scale as played by experts, $A = 459$. This unusual pitch is maintained by pipers all over the world, without the benefit of tuning forks or other aids.

The intervals are also unusual. The familiar diatonic scale has three major tones, each corresponding to a frequency ratio of 9/8, two minor tones (10/9) and two semitones (16/15). Multiply all of these together and the answer comes to

$$\frac{9^3 \times 10^2 \times 16^2}{8^3 \times 9^2 \times 15^2} = 2,$$

corresponding to an octave.

Pythagoras, better known as a virtuoso of the right-angled triangle, was also a considerable musical scholar. He knew that agreeable sounds were associated with intervals which could be expressed as the ratio of small numbers. He did not know why—nor do we. An octave (2:1) sounds fine and other simple ratios also please the ear.

The bagpipe scale uses four minor tones, one major tone and no semitones. Instead of semitones, we find two intervals of 27/25, each corresponding to a semitone (16/15), augmented by the difference between a major tone and a minor tone ($9/8 - 10/9 = 81/80$; $16/15 \times 81/80 = 27/25$). The interval 27/25 is a great limma, known to the Greeks in their mathematical doodling, but not found in any other scale today. The two limmas, between C and D and between F and G, give the bagpipe scale its distinctive quality. Simple mathematical analysis shows that the bagpipe scale provides the most efficient way of dividing an octave for playing

† Or *pibroch*, as the English call it.

pentatonic music. Most of the classical music of the bagpipe is composed in a five-note scale. The nine notes in the range (from G = 415 to high A at 918) allow the piper to play pentatonic music in three different keys: G, A and D. It is not an accident that many pipe tunes and more familiar Scots airs can be played on the black notes of the piano, which give a reasonable approximation to the pentatonic scale.

Bagpipe music is mathematically elegant and (with a little perseverance by the listener) musically satisfying too. It is a pity that more people do not appreciate it.

Left and Right

An American professor of child psychology recently reported that most mothers hold babies on their left arms—an impression reached after lengthy experiments and a tour of art galleries. The professor modestly classified his findings as 'naturalistic observations not previously reported in the scientific literature.' His search for an explanation provoked a lively correspondence (in the *Scientific American*), which called upon subliminal perception, inter-personal processes and other speculations related to the supposed need of the baby to hear his mother's heartbeat.

Writing in New York, the enquirer could hardly be blamed for gaps in his knowledge of the less popular European journals. Had he browsed further in the library he might have reached the *Proceedings of the Philosophical Society of Glasgow*, wherein Dr Andrew Buchanan, as long ago as 1862, had made luminous comments on the problem.

Buchanan studied at Glasgow University and, in 1839, was ap-pointed Regius Professor of Theory of Physic or Institutes of Medicine, a branch of the healing art which included physiology, pathology and therapeutics. He had a hard time, because in those days the Regius Profes-sors did not enjoy the superior esteem now associated with the title. On the contrary, they were regarded as interlopers foisted on the university by the government. Receiving no salary from the Crown or from the university, they had to live by private practice, and invariably survived.

Buchanan worked hard, both as a teacher and as a surgical chief in the

... before retiring (under strong pressure) at the age of 77.

Royal Infirmary, before retiring (under strong pressure) at the age of 77. Coutts, the historian of Glasgow University, delicately observed: 'towards the end of his tenure, having declined into the vale of years, he could not altogether escape the effects of age,' but, to judge from his scientific writings and his invective against the University Court, the old Professor was still in good shape.

In his first examination of the problem of left and right, he explained why most people use the right hand for delicate or complicated tasks, but the left hand for carrying things—or holding the baby.

The right hand becomes stronger, he claimed, because it is used more. But why is it used more? Because the centre of gravity of the body is not in the middle, but appreciably to the right. Although the body is roughly symmetrical, the liver (a relatively heavy organ) lies on the right side and is not balanced by the heart which, though commonly supposed to lie on the left side, has a fairly central location. The apex, where the audible heartbeat originates, lies to the left but the rest of the heart is inclined to the right.

The displacement of the centre of gravity of the body to the right of the midline has some interesting consequences. If the arms and trunk move from left to right, the centre of gravity moves still further towards the right and the stability of the body is in danger. (We stay upright only if a vertical line through the centre of gravity reaches the ground within the space bounded by the two feet. Balancing on one foot is difficult and sometimes impossible: stand at right angles to a wall with one arm and leg pressed closely against it, raise the other leg and see what happens.) But movement from right to left, as when the right arm is doing the work, brings the centre of gravity nearer to the centre of the body, so that the movement can be continued for a greater distance without the danger of overbalancing.

For carrying a heavy load, the left hand has the advantage. If a suitcase is held in the right hand, the centre of gravity of the system naturally moves further to the right and must be restored by leaning in the opposite direction. But if the load is held in the left hand, the centre of gravity moves to the left—that is, towards the midline of the body—and stability is actually improved. A very heavy load will, of course, require a compensating lean to the right. It might be thought that the baby is carried on the left arm so that the right hand is free for other tasks; but left-handed mothers also hold their babies on the left arm.

Buchanan returned to the problem in 1877, considering whether the centre of gravity of the body lies above or below the transverse axis, mid-way between the soles of the feet and the crown of the head. In men, he concluded, the centre of gravity is usually above the axis; further mechanical considerations showed that this position favours right-handedness. Women, however, being endowed with 'a most happy conformation of body,' usually have the centre of gravity on or very near the axis, giving a greater tendency to left-handedness. This conclusion has not been verified by more recent research, but Buchanan may have been right; women were a different shape in those days.

History

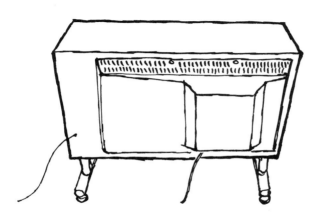

The Experiment that Didn't Work

When the new natural philosophy came to Glasgow University in 1945, its shock troops, fresh from their victories in radar and nuclear technology, joined a department bearing many outward signs of the influence of Lord Kelvin, who had been professor from 1846 to 1899.

Rearrangement of rooms brought to light many relics of the great man. Some were of little value, such as the tall U-tube filled with a muddy liquid and described rather unconvincingly as an experiment on the diffusion of liquids, begun by Kelvin a long time ago. 'We'll give it another week,' Professor P I Dee is reputed to have said, 'and if it's not finished by then, throw it out.'

Whatever the truth of this legend, a similar experiment has survived for nearly a century, and is still to be seen in the University. It consists of a small wooden staircase about 20 inches high. A block of pitch, placed on the top step a long time ago, is still hard to the touch but has, over the years, flowed like lava over the treads below.

Kelvin was fond of this experiment, which illustrated the strength and the weakness of his approach to physics. He said in Baltimore in 1884:

> It seems to me that the test of 'Do we or do we not understand a particular point in physics?' is 'Can we make a mechanical model of it?'

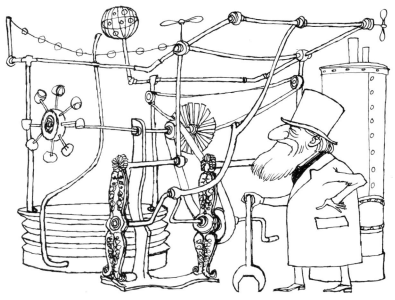

... until I can make a mechanical model of a thing.

Later in the Baltimore lectures he returned to the same point:

> I never satisfy myself until I can make a mechanical model of a thing. If I can make a mechanical model, I can understand it. As long as I cannot make a mechanical model all the way through, I cannot understand; and that is why I cannot accept the electromagnetic theory.

The electromagnetic theory of light (as developed by James Clerk Maxwell) represented a profound challenge to Kelvin's view of the universe. He believed, in the tradition of Galileo and Newton, that a mechanical explanation could be found for all of the phenomena of nature. This view of physics had great success in thermodynamics, and Kelvin pursued it vigorously in his studies of electricity and light.

By the time that he came on the scene, the wave theory of light was firmly established, having displaced Newton's belief that a beam of light was a stream of particles (or corpuscles) emitted from the lamp or other source. But if light is a wave, what is the medium through which it travels; in other words, what is it that waves about when the light passes? Sea waves displace water; sound waves are associated with vibration of the air, and seismic waves with disturbances in the Earth's crust. A light wave, of course, will move through air or water but, as evidenced by light from the Sun, will also travel quite readily through apparently empty space. To anyone who believed in a mechanical explanation of the universe, a light wave could not exist without some medium to support the undulations.

The ether was invented as the medium supporting light waves. It had remarkable properties. To carry vibrations travelling at the speed of light—186 000 miles per second—it had to be very rigid. On the other hand, since it obviously produced no retarding effect on the Earth and planets, it had to be exceedingly light.

The general idea was of a jelly-like material, with neither colour nor smell, but which was able to permeate everywhere and to quiver when light passed through it. When challenged to justify this extraordinary combination of properties, Kelvin explained that the difficulty was 'not so very insuperable.' He spoke then of the behaviour of pitch, or shoemaker's wax, a brittle solid which could nevertheless flow like a liquid, though very slowly.

The ether, said Kelvin, was a kind of wax which behaved like a solid for the rapid vibrations of light waves, but like a fluid in other circumstances. Light waves were vibrations of such high frequency that the response of the ether was perhaps hard to understand, but

> ... it is no greater mystery at all events than the shoemaker's wax.

It was obvious also that the ether must be in a state of absolute rest. Apart from the absence of friction in the solar system (the planets conform strictly to the laws of gravitation), any disturbance in the ether would be detected through changes in the light from a star as the Moon passed near it;

in fact there is no change at all, up to the moment when the Moon obscures the star completely.

It seemed that the ether, though so important, could never be detected, but an ingenious experiment made in 1887 by two American physicists, A A Michelson and E W Morley, offered some hope in this direction.

The Earth is moving at a speed of about 20 miles per second in its path around the Sun. The ether, through which light waves travel, remains at rest. Consequently, a beam of light travelling in the same direction as the Earth should move more quickly than a beam travelling in the opposite direction, just as a swimmer makes better progress with the tide than against it. The experiment that Michelson and Morley designed was sufficiently sensitive to show this effect for a light beam, and therefore to prove that the elusive ether really existed. But when the test was made, the result was astonishing.

The apparatus that they used was an interferometer; in other words, a race track for two beams of light. One beam went out to a mirror and back to the start, a total distance of about 40 feet. The other beam travelled a similar path, but in a direction at right angles to the first.

If the first beam is sent in the same direction as the Earth's motion in its orbit around the Sun, it should take a little longer to return to the starting point than the second beam, travelling in the perpendicular direction, just as a swimmer who does a mile upstream and a mile downstream will take a longer time than if he does a mile across the river and a mile back.

The two Americans looked through the telescope at the pattern of bright and dark lines which recorded the progress of the two beams of light. Then they turned the whole apparatus through a right angle and looked again to see how much the pattern had changed.

It had not changed at all. However hard they tried, they could not demonstrate any difference. The scientific community was thrown into confusion; but not for long. The obvious answer was to say that the ether did not exist. Unfortunately, the natural philosophers of those days had forgotten the story of the Emperor's new clothes and no one was ready to take the revolutionary decision demanded by the experimental findings.

A plausible explanation was offered by the Irish physicist G F FitzGerald, who suggested that the length of a moving body was reduced slightly in the direction of motion, because of interaction between the ether and the electrical forces holding the atoms together.

In the race between the two light beams, the track pointing in the direction of the Earth's motion through the ether was shortened, but the other track was unchanged. The alteration was just enough to compensate for the longer time on the up-and-down journey (as compared with the side-to-side course) and to explain the dead heat.

FitzGerald's contraction certainly explained the negative result of the Michelson–Morley experiment, but it did not survive for long. In the early years of the twentieth century, Einstein took up the problem, reflecting that an early form of relativity (proposed by Newton) asserted the impossibility of distinguishing between a state of rest and a state of uniform motion (that is, at constant speed) in a straight line.

Seated inside an aeroplane, travelling at a constant speed, there is no conclusive test which we can perform to tell whether the aircraft is moving past a stationary cloud or whether the cloud is moving in the opposite direction past a stationary aircraft. In a more general way, all of the laws of mechanics are exactly the same in the aircraft as they would be on the ground; the coffee comes out of the pot at the same rate, and the voice from the loudspeaker sounds just as distorted.

Einstein disposed of the Michelson–Morley experiment without using FitzGerald's convenient contraction. No one disputed Newton's conclusion that any experiment in mechanics would give the same answer whether the laboratory was at rest or whether it was moving steadily in a straight line. Yet Michelson and Morley were suggesting that an experiment in optics would show a different result. This was absurd, thought Einstein. The laws of nature must be the same whether we study them in the optics laboratory or in the dynamics laboratory, and therefore there cannot be any ether.

These conclusions ... published in 1905, marked the end of the ether.

These conclusions formed the basis of the Special Theory of Relativity, published in 1905, and marked the end of the ether, which kept turning up occasionally in textbooks for another 30 years, but finally came to rest with caloric and phlogiston among the concepts which were useful while they lasted, but are now better forgotten.

Einstein's theory of relativity seems to us today a demonstration of the power of common sense, but when it appeared it provoked a lot of controversy. It was not until 1922 that the Swedish Academy of Science decided to give him the Nobel Prize, and even then they avoided mentioning relativity in the citation.

Having disposed of the ether and restored a pleasing uniformity to the laws of science, Einstein pressed on and explored the consequences of the new principle. He found that it led to a fresh formulation of the basic laws of mechanics. Newton's laws of motion were accurate for slowly moving bodies but, in general, the mass of an object increased as its speed increased.

The change is insignificant for a train or a motor car, but the mass of an electron may be doubled as it shoots across the space inside an x-ray tube. Having shown that the mass of an object increased when its energy increased, Einstein went on to prove that the mass of a body would be reduced if energy was given off. In short, mass and energy were equivalent.

The rate of exchange was enormous, expressed in the now-familiar equation: $E = mc^2$, in which E represents energy, m is mass, and c is the speed of light. In practical terms, one gram of any substance is equivalent to 25 000 000 kilowatt-hours of energy.

This extension of the relativity theory solved a baffling problem by providing an explanation for the continuing radiation from the Sun, which is now known to depend on the slow but fruitful conversion of matter into energy. The same simple process underlies nuclear power and the atomic bomb. A lot of history clings to Kelvin's toy staircase.

Forefathers of Flight

The prospect of flying through the air, which inspired the earliest writers of science fiction, also encouraged engineers and charlatans for several centuries before the success of the Montgolfier brothers in 1783.

The hot-air balloon needed no technology that had not been available a hundred years earlier. Indeed, a serious attempt to build a passenger-carrying balloon was made early in the fifteenth century.

This experiment failed because the designer (an Italian whose name has not survived) did not build big enough. In the days before physics and chemistry became exact sciences, the would-be aeronaut did not appreciate that the difference in density between hot and cold air is quite small. Even in the eighteenth century, designers thought about the composition of the smoke rather than about its temperature, which is all that matters; the Montgolfiers tried various combustibles (including old shoes and well hung meat) before settling for paper, wool and straw.

The major concern of the early inventors was with man-powered flight. Here also their technical judgment was faulty. They were nearly right in believing that human muscle power alone would not lift a man into the air for sustained flight; even today this is almost impossible. The problem of ascent was avoided by starting from the top of a cliff or tower; but the outcome was almost always disastrous.

Eilmer of Malmesbury, a Benedictine monk of the eleventh century, was reliably reported to have flown for more than 200 yards before making a violent landing which broke his legs. With remarkable insight he blamed his instability on his failure to provide a tail to supplement his wings.

When John Damian, the Abbot of Tungland, launched himself from the battlements of Stirling Castle on 27 September 1507, he hoped to

... he blamed his instability on his failure to provide a tail.

make a non-stop flight to Paris in order to demonstrate the power of alchemy to his employer, King James IV of Scotland.

Damian's flying apparatus was modelled, not surprisingly, on the wings of a bird, but it did not work very efficiently for, as William Dunbar explained in a satirical poem:

> He schewre his feddreme that was schene
> And slippit out of it full clene
> And in a myre, up to the ene
> Amang the glar did glyd†.

John Damian ... launched himself from the battlements of Stirling Castle.

The accident was described more precisely by Bishop Lesley:

> He flew of the castell wall of Striveling, bot shortlie he fell to the ground and brak his thee bane.

Damian was fortunate to escape with nothing worse than a broken thigh bone. He did not understand the aerodynamics of flight as demonstrated by the birds; nor did anyone else until more than four centuries later.

The pioneers were impressed by the effortless skill of the birds and so, understandably, concentrated their efforts on making wings. Leonardo da Vinci guessed, quite correctly, that the lifting force depends on a difference in air pressure above and below the wing—but he believed (mistakenly) that the effect was created by the beating action which compressed the air under the wing, 'and the pressure from this air raises up the bird.'

None of the early inventors appreciated that mere flapping of the wings was not enough to sustain flight. A bird, like an aeroplane, needs lift to keep it above the ground and thrust to move it forward through the air.

† Schewre = tore off; schene = bright; glar = mud.

Lift is obtained in much the same way by both a bird and an aeroplane. As the wing moves forward the air stream is divided, some flowing under and some over the wing surface.

The upper path is longer and the air therefore has to travel faster to keep up with the other half which has taken the low road. When the air is travelling faster it exerts a lower pressure. This effect (used also in the scent spray or atomiser) leads to lower air pressure on the upper surface than on the lower surface of the wing. Consequently, the wing is able to remain in mid-air.

There is, of course, another large problem in securing a rapid flow of air over the wings. The propeller of an aeroplane is really a kind of wing, giving horizontal thrust rather than vertical lift. A bird uses the large feathers near the tip of the wings to do the job of the propeller.

During flight these feathers move in a figure-of-eight pattern, twisting and bending continually to adjust the direction and pressure of the air in a way which gives thrust as well as lift. This method works quite well and gives the bird a remarkable delicacy of manoeuvre, but it is certainly not as efficient as the rotary motion of a propeller blade.

Birds solved the problems of swing-wing geometry and retractable undercarriages long ago. In a less obvious way they have achieved remarkable success in weight reduction. One bird with a 7 foot wing span was reported to have a skeleton weighing only 4 ounces. A bird's head is very light, because it has no teeth and consequently saves considerably on jaw bones and accompanying muscles. The gizzard (which does most of the grinding and mixing that would otherwise be a job for the teeth) is relatively heavy but is located fairly well back in the body, so that the centre of gravity is behind the centre of lift associated with the wings.

This means that a bird is inherently unstable in flight, but the instability is advantageous in many ways, because it makes the structure much more manoeuvrable. An automatic pilot, such as the bird's tiny brain provides, can give effective control in this situation, but an aircraft designer would not take the risk, since the human pilot would be helpless if the automatic navigation system failed.

A bird's engines are quite efficient, using only premium fuel such as seeds, worms, insects and other high-protein foods, but very little roughage. A bird's digestive processes are rapid, sometimes taking only a matter of minutes, and a high proportion of the food intake is converted into body weight.

A bird's blood circulation is also arranged on frugal lines. Hens and turkeys have quite a good blood supply to the legs, since they need to run about, but very little to the breast and wing muscles, thereby providing the gourmet with the agreeable choice between dark meat and white meat. Although the lungs are small, they connect with several air spaces in the bones and elsewhere in the body. In this way, a bird obtains an adequate cooling capacity and also has the equivalent of a supercharger when additional effort is needed for rapid movement or manoeuvre.

Obsessed by the desire to imitate birds, the inventors of medieval and Renaissance times ignored the possibilities suggested by the windmill, which was the forerunner of the aircraft propeller. A toy familiar to modern children consists of a four-bladed propeller on a vertical stick. If the propeller is spun fast enough, by a string or screw, it takes off and flies through the air. Toys of somewhat similar appearance are shown in fifteenth-century manuscripts, but the propeller is never shown in flight, so the idea may not have been fully exploited.

Leonardo designed a helicopter, and a small model may actually have flown, but he did not pursue the idea. Of course there is no example of rotary motion anywhere in the living world, and it is perhaps not surprising that effort was directed towards the imitation of nature.

The early engineers were often sufficiently learned in other disciplines to speculate on the long-term effects of their experiments. Francesco da Lana, a seventeenth-century Jesuit, discovered the principle of the hot-air balloon, but abandoned his research because he believed that success would be dangerous to the peace of the world!

Over the Waves

An eminent lady wrote to me some time ago asking for information about 'the modest Scottish scientist who helped Marconi discover wireless.'

She gave a few clues, which I have not had the opportunity of pursuing. The trail leads to a pond near High Wycombe, where a model submarine was controlled by radio; the inventor emigrated and is believed to have died during the 1960s.

Certainly there were a few people who might have claimed priority over Marconi; some of them actually demonstrated wireless communication without recognising its significance—or potential profitability. Marconi himself wrote:

> ... the name of James Bowman Lindsay must go down to posterity as that of the first man who thoroughly believed in the possibility of long-distance wireless telegraphy.

Lindsay studied at St Andrews and taught in Dundee, when not occupied with his *Pentecontaglossal Dictionary*—a philological study in 50 languages, which was to reveal the time and place of man's origin—or with his electrical inventions. In 1845 he published a scheme for an Atlantic cable, and in 1853 the *Dundee Advertiser* announced that

> ... he has now discovered that instantaneous intelligence can be transmitted to all parts of the world without the aid of a submerged wire.

Lindsay's idea was to use the water itself as a conductor. The transmitter consisted of a battery and a telegraph key connected to two large plates immersed in water a few hundred yards apart. Two similar plates near the opposite bank of the River Tay were joined to a receiver. The trials were quite successful and the technique might be described as wireless telegraphy, but the range was limited to a few miles and the idea was not developed.

The first person to experiment with radio waves was probably Galvani, who noticed (in 1780) that sparks from an electrostatic machine caused a dead frog to jump, even though it was several feet away. In 1842 Joseph Henry (a professor at Princeton who had already invented the electric telegraph and the electric motor) found that a spark would magnetise needles ten yards away. He attributed this effect to an electrical disturbance travelling through the intervening space, which was the correct explanation. A spark is a good source of radio waves, and was used in Marconi's early transmitters.

Nearly a century after Galvani, Edison noticed that sparks could be drawn from any metal object in the vicinity of a working magnetic vibrator (of the kind still used in electric bells), even though there was no connection with the vibrator. He thought that a new 'etheric force' was responsible. Elihu Thomson (a lad of 18 who afterwards helped to found the Thomson-Houston Company) observed the same effect with an induction coil.

Both he and Edison afterwards regretted their inability to see the significance of the experiments. 'What has puzzled me ever since is that I did not think of using the results of my experiments on etheric force,' Edison wrote later. 'If I had made use of my own work I should have had long-distance telegraphy.'

Amos Dolbear, an American teacher, demonstrated a new system of telegraphy in London in 1882. The transmitter (including a microphone) was connected between a gilded kite and a metal plate buried in the ground. The receiver was connected between a tin roof and the ground, but clear signals (including renderings of 'Yankee Doodle' and 'God Save the Queen') were heard when the receiving apparatus was merely held in the hand, unconnected to anything.

As early as 1879, David Hughes walked up and down Great Portland Street in London with a telephone to his ear, listening to noises produced by an induction coil in his house a few hundred yards away. Three Fellows of the Royal Society observed the effect and discouraged him from proceeding further. Although Edison and some of the other early experimenters were not learned in natural philosophy, the delegation from the Royal Society ought to have linked Hughes's demonstration with the theoretical prediction of radio waves made by Maxwell in 1864.

Three Fellows of the Royal Society ... discouraged him from proceeding further.

In the event, it was the early death of Heinrich Hertz that led to the decisive advance. Hertz, a professor at Karlsruhe, was only 30 when he established that the electromagnetic waves predicted by Maxwell could be produced in the laboratory and detected at a distance.

He died in 1894, only 36 years old. Oliver Lodge, the English physicist (and amateur of the occult), gave a commemorative lecture, presenting his own work in a favourable light. The lecture was published and came to the attention of Marconi, then 20 years old, and inspired him with the idea of sending messages by the new waves. Alexander Popov also had the same idea and, in his native Russia, is often credited with the invention of radio.

Some of the early workers were practical men who did not know enough science to bridge the gap between experiment and theory. Some were diverted by accepting wrong explanations, and one made the excusable mistake of listening to the Establishment. There is a lesson for inventors in this story: most of you will never make a penny but one of you may be sitting on a gold mine.

Old Flame

An English chemist was reported a few years ago to have discovered the secret of Greek fire. It is not much of a secret (Professor J R Partington wrote a book about it in 1960), but the story is certainly interesting.

Terrifying weapons—both real and imaginary—were as important in ancient times as they are today. Some of them were very effective, including early versions of the flame-thrower and the Molotov cocktail.

The first incendiary devices were based on Middle East oil which, though often supposed to be a recent discovery, has been around for a very long time. Various forms of petroleum seep through the ground to appear on the surface at places in Iraq and Iran. Crude oil was imported into Constantinople well over a thousand years ago, and was used for heating public baths.

Crude oil was imported ... and used for heating public baths.

Lighter oil products were obtained by distillation, a process invented by the first woman scientist. This was Mary the Jewess, an Egyptian alchemist of the early Christian era. She also invented the water-bath, still known in France as the bain-marie.

The *Liber Ignium ad Comburendos Hostes*, or 'Book of Fires for Burning Enemies' (written in the twelfth century, but based on earlier Greek sources), describes various forms of self-igniting firelighters. A powerful mixture of sulphur, bitumen, pyrites, mulberry juice and quicklime was to be stored in airtight boxes. 'If you wish to set fire to the arms of the enemy, secretly spread over them this preparation at night. When the sun rises all will be burned.'

Livy tells that, during the Bacchanalia of 186 BC, some of the swingers carried torches made of sulphur and quicklime which burst into flames when dipped in water. A Chinese historian, describing a twelfth-century naval battle, tells of missiles containing sulphur and quicklime in paper bags, which caught fire when they hit the water. These devices are almost certainly legendary, but Greek fire was genuine enough.

The name is misleading, for the weapon was invented and used successfully for centuries in Constantinople. It seems to have appeared at about the end of the seventh century and to have helped significantly in beating back the invasions which continued until the city finally fell to the Turks in 1453.

Greek fire was simply burning oil, propelled by a primitive stirrup pump. A contemporary chronicler writes of 'small siphons discharged by hand from behind iron shields ... these can throw the prepared fire into the faces of the enemy.' A later writer describes a syringe used in surgery as 'like the tube by means of which naphtha is thrown in sea-combats.'

The Arabs, after being singed a few times in attacking Constantinople, found how to make Greek fire for themselves and trained special naphtha troops, protected by flame-proof clothing. The Arabs also made a 'fierce fiery oil' which was probably a lighter fraction obtained—like petrol—by distilling the crude materials.

The Crusaders were disconcerted by Greek fire, which had been used against them in many sieges. Psychological warfare was also quite effective. The Byzantine chemists put about the story that the secret had been revealed to the Emperor Constantine by an angel and that anyone disclosing the details would die in the act.

It was also widely believed that troops not burned to death would succumb to the lethal fall-out which followed the flames. The fiery oil,

... the secret had been revealed to the Emperor Constantine by an angel.

hurled in bottles, was reputed to destroy castles with a loud noise and a terrifying whistle.

Historians of crusading times relate that Greek fire used a liquid obtained 'from springs in the East.' Distilled naphtha (or petrol) would not have carried very far, but the ancient recipes suggest that the flammable mixture was thickened with sulphur and resins, allowing it to travel further and to stick to the target.

Greek fire was a sophisticated weapon which has survived, with relatively minor improvements, into the twentieth century. The skill of the ancient military engineers need not surprise us, for technology was well advanced in early times. The Romans had concrete, piped water and central heating. Plato had a water-driven alarm clock, which blew a whistle at dawn to signal the first lecture of the day; other Greek inventors anticipated the cuckoo clock and the timepiece with a built-in calendar.

The alchemists, though they are often supposed to have been charlatans or confidence tricksters, were skilful technologists. Greek fire was one of the many genuine achievements of the ancient engineers, who can still teach us useful lessons.

How Science Doesn't Happen

DNA is a substance which holds significant information about heredity and which, in a sense, contains the secrets of life itself.

Its molecules have a complicated structure which was studied for many years by the well established methods of x-ray analysis. Pictures obtained by Maurice Wilkins and Rosalind Franklin at King's College, London, in the early 1950s suggested a helical structure, which was verified in an ingenious and convincing way by James D Watson and Francis Crick, working at the Cavendish Laboratory, Cambridge.

Out of this important story, Honest Jim Watson (the title was conferred by one of his former associates) has constructed a rattling adventure yarn full of suspense and spies, with a vigorous cast of goodies and baddies†.

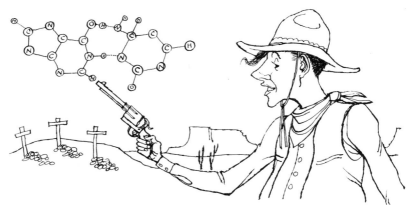

Honest Jim Watson ... with a vigorous cast of goodies and baddies.

There is only one goody—Dr Watson himself—but a rich variety of baddies, including his chief collaborator ('I have never seen Francis Crick in a modest mood .... Already for 35 years he had not stopped talking and almost nothing of fundamental value had emerged'); his principal benefactor ('Clearly Rosy had to go or be put in her place ... unfortunately Maurice could not see any decent way to give Rosy the boot'); a crafty American rival ('Cal Tech's fabulous chemist Linus Pauling was not subject to the confines of British fair play'); a man with a secret ('The painful fact that the picture belonged to Maurice could not be avoided. There was nothing else

† James D Watson 1968 *The Double Helix* (London: Weidenfeld and Nicolson).

to do but talk to him'); and the Cavendish Professor of Experimental Physics ('The thought never occurred to me then that later on I would have contact with this apparent curiosity of the past ... our professor was completely in the dark about what the initials DNA stood for').

Lord Snow's commendation, printed on the jacket of *The Double Helix*, describes Watson's book as 'Like nothing else in literature'—a judgment which need not be disputed—and goes on to say '... it gives one the feel of how creative science really happens.'

Creative science, it seems, is not pursued as a service to the community nor as the fulfilment of intellectual curiosity. It is practised for the purpose of winning a Nobel Prize, pushing everyone else out of the way if necessary. Confidential Medical Research Council reports and private letters (from Pauling to his son, a Cambridge student) were scanned for news of progress in rival laboratories, but news of the Cambridge work was closely guarded ('Keeping King's in the dark made sense until exact co-ordinates had been obtained for all the atoms ... Pauling first heard about the double helix from Delbruck ... I had asked that he not tell Linus ... my request however was ignored').

The President and Corporation of Harvard explained, after deciding that the University Press should not publish Watson's book, that they did not wish to become involved in a dispute among scientists. They might have added that the text (though mercifully short) displays no literary merit. The story is told as a string of trivial anecdotes, interspersed with gobbets of biochemical jargon and schoolboy slang: 'Lying low made sense because we were up the creek with models based on sugar–phosphate cores. No matter how we looked at them, they smelled bad.'

Every book is a picture of the world as the author sees it. Watson describes a realm of fantasy, dominated by greed, rudeness, and chicanery, as close to the real world of science as Batman is to Leonardo. *The Double Helix* is certainly a literary curiosity, fit to be placed alongside *The Young Visiters* and *Delina Delaney*, but as a contribution to the history of ideas or the understanding of science it is not significant.

Jodrell Bank's Telescope

The familiar outline of the Jodrell Bank radio telescope has become a symbol of Britain's scientific prowess. In the technical sense, it is certainly a remarkable achievement; the steel bowl which forms a reflector for the transmitting or receiving aerial is 250 feet in diameter, yet can be electrically driven to point anywhere in the sky.

In the scientific sense, Jodrell Bank has been overshadowed by Cambridge, where the Mullard Radio Astronomy Observatory commands equipment of greater sensitivity. The main interest of Sir Bernard Lovell's story† is, however, in the administrative misfortunes which he describes and in their political implications.

The Jodrell Bank project went wrong because large sums of money were spent without proper authority. Lovell's original estimate of £50 000, made in 1949, was quickly overspent as the design was studied in more detail, and the cost had reached £335 000 by March 1952, when it was announced that the bill would be met by grants from the Department of Scientific and Industrial Research and the Nuffield Foundation.

Not surprisingly, the costs continued to rise from the day the digging began. No one had ever built such a machine before. Many of the cost calculations, though quoted as firm estimates to the DSIR, turned out to be rough guesses—and were always too optimistic. Lovell, impatient over delays in construction, proposed or agreed to design changes which added greatly to the cost but were not reported to the authorities financing the project.

More overspending was not the main cause of the subsequent troubles. Disappointing amounts of public money are regularly wasted on aircraft, missiles, and army boots without much public or official concern. Unfortunately, Lovell's overspending was on the wrong sort of project and at the wrong time. The bill for the radio telescope eventually rose to £650 000. Some of the excess was defrayed by a public appeal, and some by Lord Nuffield, but the rest had to come out of the taxpayer's pocket.

The Public Accounts Committee, already suspicious of the universities' freedom from financial control, were confused and embarrassed, almost to the point of humiliation, in their efforts to investigate the Jodrell Bank situation.

The enquiries crystallised the misgivings and anxieties which had been rumbling for several years, and contributed significantly to the situation today, in which the universities are obliged to open their books and records to the scrutiny of the Comptroller and Auditor General.

Lovell's financial troubles had a variety of causes, but the Public Accounts Committee were particularly concerned about a substantial change in the design of the telescope made in 1952, when it became known that

† Bernard Lovell 1965 *The Story of Jodrell Bank* (London: Oxford University Press).

distinctive radio signals were likely to come from the Milky Way at a wavelength below the limit previously agreed for efficient operation. To cope with the new situation the two-inch square mesh forming the reflecting surface of the 250 foot bowl had to be changed to solid plate.

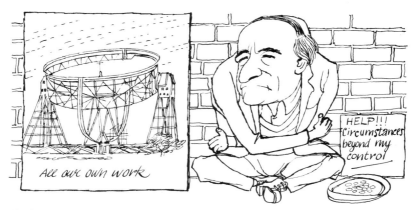

Jodrell Bank's telescope.

In October 1955, Manchester University (to whom the grants had been made) told the Department of Scientific and Industrial Research that the design of the telescope had been changed substantially without their approval. Questions asked by members of the committee and answered by the Secretary of the Department of Scientific and Industrial Research made the matter clear.

Q: If this was a university project, who had changed the design without obtaining the approval of the university, the engineer or Professor Lovell?

A: Oh, no, quite clearly the engineering consultant changed the design without the concurrence of the university.

Q: He designed that on his own without consulting anybody?

A: Yes.

Q: And without advising them that it would materially increase the cost?

A: Yes.

Not surprisingly, the committee reported that unauthorised changes made by the consulting engineer (Dr H C Husband) had contributed significantly to the increased cost. Husband was very angry and asked Lovell to write a letter to *The Times* denying these allegations, which he knew to be untrue. Lovell did not feel able to help him:

> I replied that the PAC report was a privileged document and that I could not possibly move without the permission of the Vice-Chancellor. He was on holiday ...

Husband then wrote to the Department of Scientific and Industrial Research setting out the facts which, for reasons that defy analysis, the Department had not taken the trouble to discover before giving evidence. The Public Accounts Committee returned to the matter and reversed their decision:

> It is clear that the evidence given to the committee of last session was gravely inaccurate and misleading, and that there was in fact the fullest collaboration on scientific and technical matters between the consultants and the university Professor.

Eventually the commotion subsided and the telescope was paid for. After tracking the rocket which carried the first Sputnik, the Jodrell Bank equipment helped in several studies on American and Russian Moon probes, gaining well deserved public applause, even if the work was not of great scientific value.

Lovell rightly insists that the telescope, even at its final price, was a good bargain and that it could not have been built for less. His picture of himself as the unworldly scientist, bewildered by financial and engineering problems, baffled by official ingratitude, and prostrated with anxiety lest innocent miscalculation should land him in jail, is well drawn, even if not utterly convincing. His book, fluently written with abundant quotations from diaries and other contemporary documents, provides a fascinating commentary on the interplay of science, politics and bureaucracy.

Finding Time

Science is still regarded in Britain as an activity for scholars and as the source of useful or profitable technology. Give us another accelerator, another computer, another laboratory (say the scientists) and you will eventually see some benefits; don't ask us what, just hand over the money and trust in science to deliver the goods.

This attitude, the modern version of the alchemists' prospectus, still convinces politicians, although some costly ventures of recent years have produced nothing at all. Its persistence is surprising, since so much of science springs from technology, which is itself the response to questions posed by the humdrum activities of people going about their daily work.

Astronomy, now one of the most recondite of the sciences, grew largely out of a single problem: the mariner's need to know his position at sea. The Royal Observatory at Greenwich and the office of the Astronomer Royal were established to tackle this problem, which also inspired Newton's work leading to the theory of gravitation, the cornerstone of modern science.

Coastal navigation needs good charts and sharp eyes, but on the ocean the task is different. Once beyond sight of land, the sailor can establish his position only if he knows his latitude and longitude. The measurement of latitude is not very difficult, given simple instruments, a little knowledge of the stars and a book of astronomical tables. Chaucer gave a good account of the astrolabe, an early instrument for measuring the altitude of stars. A sea captain is mentioned in *The Canterbury Tales*:

> He knew wel alle the heavens, as they were
> From Gootland to the Cape of Finistere.

Well! It's certainly not Madeira

Columbus did not know whether he was approaching Madeira or the Azores.

Early mariners tried to fix their position on the ocean by dead-reckoning, based on the estimates of the ship's speed and direction since leaving port. The ship's log was a piece of wood thrown overboard; knowing the length of the ship, the speed could be calculated from the time taken to pass the log. The errors in this process were enormous. On one voyage, Columbus did not know whether he was approaching Madeira or the Azores.

Measuring longitude is really a matter of measuring time. Since the Earth turns through 360° in 24 hours, the time changes by one hour for every 15 degrees of longitude. It is only by convention (reinforced by legislation in 1880) that uniform time is used throughout the United Kingdom. Until well into the nineteenth century, Birmingham was still eight minutes behind London, and Yarmouth was eight minutes ahead. These differences did not matter much in the days of the stagecoach, but became very inconvenient with the development of more rapid transport; the Post Office and the railway companies encouraged the change.

A mariner can readily find his local time; for example, the Sun is (to an observer in the northern hemisphere) due south at noon. If he knows what the clock is then showing at his home port, he can easily calculate the difference in longitude. Today his radio gives time signals, but 400 years ago the problem was more difficult. In 1530, the Flemish astronomer Gemma Frizius suggested that ships should carry an accurate clock, registering the time at the port of embarkation. The idea was simple but optimistic. Nearly 200 years later, Isaac Newton observed:

> By reason of the motion of a ship, the variation of heat and cold, wet and dry, and the difference in gravity in different latitudes, such a watch hath not yet been made.

Lacking good clocks, sailors and scientists looked upwards. The Sun, Moon and stars form a celestial clock with the sky as its dial. Unfortunately, their wanderings are complicated and not easily predicted. But if the motion of the Moon and stars can be calculated for one place on the Earth (for example, Greenwich), the problem is solved. A mariner can observe the positions of suitable stars relative to the Moon, note the local time and, from the *Nautical Almanac*, find the time at which the stars would occupy the same positions when viewed from Greenwich. The difference in time is easily converted to the difference in longitude.

Many great astronomers (including Galileo and Newton) tried, with little success, to predict the future motion of the Moon. Several governments offered prizes for a solution to the problem of finding the longitude. The largest prize, of £20 000, was offered by the British Government and was eventually paid with great reluctance.

The solution came, not from navigators or astronomers, but from John Harrison, a Yorkshire carpenter who built clocks of astonishing accuracy. His best chronometer, made in 1761, was not much bigger than a pocket watch, and kept time to within a tenth of a second per day.

Unfortunately, the inventor had to wait a long time for his prize; it was not paid until George III intervened in 1772. The *Nautical Almanac* first appeared in 1767, and by that time the major problem of navigation had been solved.

Accident or Plan?

Is Man a chemical machine—or is he governed by principles and powers that transcend the man-made laws of science? Is evolution the result of blind chance—or is it the expression of a basic property inherent in the make-up of the universe?

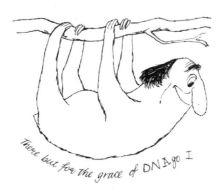

There but for the grace of DNA go I

Is evolution the result of blind chance?

These are not new questions, but every generation of scientists tackles them afresh, confident that currently fashionable discoveries and ideas will reveal the truth by dispelling the misconceptions and prejudices of the past.

Jacques Monod writes† tersely but fluently, expressing the views of the biochemists who have now inherited the mantle of omniscience from the physicists. He begins by asking: What is life? Living creatures have the ability to reproduce themselves, a process requiring the accurate coding and transmission of much complex information on structure and function.

The property of unvarying replication (later realised to be modified by evolution) has inspired arguments among moral and natural philosophers for more than three hundred years. Science, says Monod, depends on the proposition that nature is objective; that 'true knowledge cannot be reached by interpreting the phenomena of nature in terms of final causes—that is to say, of "purpose".' But living organisms do appear to be built and to act with a purpose. This paradox, says Monod, provides the central problem of biology.

Reviewing earlier attempts to resolve the contradiction, he classifies the seekers as

† Jacques Monod 1972 *Chance and Necessity* (London: Collins).

animists or vitalists, and dismisses them all: Leibnitz, Hegel, Bergson, Marx, Engels and (descending to the lightweights) Teilhard. Vitalism teaches that living creatures are radically different from inanimate things and are not governed by the same physical or chemical laws. Animism sees evolution and purpose at work throughout the cosmos, living or non-living, with Man as the ultimate and highest product.

Bergson's *elan vital*, distinguishing living matter from stones; the Marxist's attempt 'to systematise a subjective interpretation of nature, showing it to have an ascending constructive creative intent, a purpose'; and Teilhard's ascending energy: all deny the objectivity of science which is the foundation of Monod's philosophy.

A large part of his book is devoted to the biochemical alternative to animism. We now know, he explains, that the chemical machinery of living matter is essentially the same, from bacterium to Man. More importantly, the only source of change is the random and accidental rearrangement of molecules in the nucleic acid structures which carry the genetic message. Evolution depends on blind chance. This proposition is advanced, not as a speculation to be discussed, but as the only conceivably valid hypothesis 'compatible with observed and tested fact.' It will, says Monod, never be revised.

It goes without saying that this is an anti-religious book. Man emerges from the scientific analysis as a helpless being 'alone in the unfeeling immensity of the universe,' owing nothing to God's purpose or to his own efforts, but created out of an accidental tangle in a protein molecule.

Although rich in fascinating expositions of some of the most important topics in contemporary science, Monod's book is essentially anti-scientific too. There have been times when science seemed capable of explaining everything, and moments when baffling complexities were suddenly dissolved through the discovery of unifying principles.

But the laws of science are not a jigsaw waiting for the last piece to be added. They are laws made by Man as the distillation of experience, and there is nothing immutable about them. The genetic code is a great achievement of hand and mind fit to rank with relativity, radioactivity, electricity, and the other major achievements of science—but it is not the last word.

The Cavendish

Britain's greatest scientific achievement is just over 100 years old. When the Cavendish Laboratory opened its doors in 1874, English science was at a low ebb—nowhere more noticeably than in Cambridge. Since then, the Cavendish has dominated the world of physics—and sometimes the whole world of science—through a series of spectacular discoveries, usually achieved with simple apparatus in humble surroundings. Electronics was born in Cambridge, with J J Thomson's discovery of the electron. The nuclear age rests on work done in the Cavendish by Rutherford and his team, including the discovery of the neutron—the critical factor in the development of atomic energy and nuclear weapons. More recently, the Laboratory has made decisive advances in molecular biology (leading to the discovery of the double helix and the genetic code), and in radio astronomy.

Cambridge took a long time to adapt to the intellectual shock delivered by Isaac Newton. Even 150 years later, the University was still primarily a place for the education of Anglican clergymen; Newton himself had originally gone there for that purpose. As Mr J G Crowther says, in his history of the Cavendish Laboratory[†], 'Newton's works were greatly esteemed but were taught rather as the theologians taught the Bible.' Meanwhile, however, new ideas were germinating elsewhere. In Glasgow, William Thomson (afterwards Lord Kelvin) started teaching physics as an experimental subject in the 1850s; the same approach was taken up in Berlin, Oxford and Manchester during the 1860s. Cambridge had established the natural sciences tripos in 1851, but physics was still being taught as a branch of mathematics. The existing professors were unwilling for any extension of their traditional (and sometimes modest) duties; one of them observed that 'it can hardly be expected that he should now take up a course of lectures on subjects to which he has not paid special attention.'

The necessary reform was brought about largely by the Duke of Devonshire, who succeeded the Prince Consort as Chancellor of the University in 1861. He had a considerable interest in science and was familiar with the work of his kinsman, the Honourable Henry Cavendish, the eminent eighteenth-century amateur of science.

At that time there were hardly any professionals; the people who kept physics, chemistry and natural history alive were doctors, clergymen, soldiers or—lacking any other excuse—just plain rich. Cavendish belonged to the last category. His grandfathers were both Dukes (of Kent and Devonshire) and he treated an immense fortune with splendid indifference. His banker, finding a balance of £80 000 in the current account, called one day to advise a few investments. 'If it is any trouble to you,' said Cavendish, 'I will take it out of your hands, so do not come here to plague me.' Cavendish was the first to apply highly accurate methods of measurement in

---

[†] J G Crowther 1974 *The Cavendish Laboratory 1874–1974* (London: Macmillan).

scientific research, setting standards which are still valid. As one biographer observed, he regarded the universe as 'a multitude of objects which could be weighed, numbered and measured; and the vocation to which he considered himself called, was to weigh, number and measure as many of these objects as his allotted three score years and ten would permit.'

'a multitude of objects which could be weighed, numbered and measured.'

At one end of the scale he weighed the Earth; at the other, he discovered the composition of water and made some penetrating observations which anticipated the atomic theory developed early in the nineteenth century. Cavendish died in 1810, leaving more than a million pounds to his relatives; some of this fortune came back to the world of science with spectacular results.

When Cambridge University decided in 1871 to establish a professorship of experimental physics, the Duke of Devonshire offered to meet the expense of building and equipping a laboratory; the cost was £8450. In the search for a professor, the University turned first to Sir William Thomson, then the most eminent British physicist. He did not wish to leave Glasgow, where he had a thriving laboratory and lucrative commercial interests. The choice fell on James Clerk Maxwell, the Laird of Middlebie, Dumfriesshire, in Scotland. Maxwell was no stranger to the academic life, having been appointed at the age of 24 to the Chair of Natural Philosophy at Marischal College, one of the two universities then flourishing in Aberdeen. He had no great ambition for high office and wrote to his father (from Trinity College, Cambridge where he was a Fellow): 'I think the sooner I get into regular work the better.' His father replied, 'I believe there is some salary, but fees and pupils, I think, cannot be very plenty. But if the postie be gotten, and prove not good, it can be given up; at any rate it occupies but half the year.'

Maxwell stayed in Aberdeen for only three years. When the two universities were amalgamated in 1860, there was room for only one professor in natural philosophy, and Maxwell was made redundant. He returned to the farm but soon afterwards became Professor of Physics at King's College, London, where he remained for five years before resigning through ill health.

Maxwell's work was almost entirely mathematical. He predicted the existence of radio waves and accurately calculated the speed of light, long before it was measured, but when he went to Cambridge in 1871 he responded to the spirit of experimental enquiry which was the inspiration of the new Laboratory. The building work took three years, during which time Maxwell lectured where he could. In 1872 he wrote: 'I have no place to erect my chair, but move about like the cuckoo, depositing my notions in the chemical lecture room first term; in the botanical in Lent, and in comparative anatomy in Easter.'

The new institution was generally known as the Devonshire Laboratory. During the opening ceremony R C Jebb, the Public Orator, delivered a Latin oration, thanking the Duke for his generosity and suggesting that the Laboratory should be given the name of the Cavendish family; the Duke agreed, also in Latin. Maxwell was able to report that the Laboratory 'contained all the instruments required by the present state of science.' As one of his successors remarked, many years later, 'it has never since been in this condition.'

When Maxwell died in 1879, his successor was Lord Rayleigh, a gifted amateur who preferred the life of a country gentleman to that of a university teacher. He was, however, in some financial difficulty; the agricultural boom in the United States during and after the Civil War produced surplus wheat which was exported to Britain at prices which British farmers could not match. By 1884, the position had improved and Rayleigh resigned the chair to return to his estate in Essex. As Crowther observes:

> Maxwell, the Scottish laird, and Rayleigh, the English landed aristocrat, were detached in their attitude to the chair and direction of the Laboratory. If they had been better off financially at the time they might not have accepted them. The independence of a gentleman was still socially more important than the most eminent scientific post.

When the electors met to select a new professor, their first choice was Sir William Thomson, as it had been in 1871 and 1879. He was again unwilling to move and, at 60, was too old anyway. The successful applicant was J J Thomson, a young man of 28.

Thomson reigned for 34 years and was a little reluctant to give up the direction of the Laboratory, even after his appointment as Master of Trinity College in 1918.

For some time after 1884, the Cavendish Laboratory continued to be the centre of English physics. An important change was made in 1895 when the University agreed that graduates of other universities could be admitted as research students. The first to be admitted under the new regulation was an Ernest Rutherford from New Zealand; after periods in Montreal and Manchester, he succeeded Thomson in 1918.

J J Thomson's discovery of the electron (in 1897) was the key to radio, television, electronics, computers and much else in twentieth-century technology. His successor opened the door to the nuclear age. When Rutherford took over the Cavendish Laboratory he had already discovered the atomic nucleus and the general principles on which atoms are put together. In 1920 he predicted the existence of the neutron and forecast many of its properties. The team grew, as brilliant students arrived from all parts of the world, but the search for the neutron was long and frustrating. By 1930, nuclear physics in Cambridge seemed to be approaching a dead end. Rutherford himself was nearly 60; some of his collaborators had left the subject, uncertain of its future; and those who remained were struggling with experiments of ever-increasing complexity in severely overcrowded conditions.

The clouds lifted in 1932, when Chadwick discovered the neutron and, almost at the same time, Cockcroft and Walton disintegrated the atomic nucleus by bombardment with high-speed particles produced in an electrical discharge tube. This experiment gave the first direct proof of the conversion of mass into energy, which had been predicted by Einstein nearly 30 years earlier—a process which forms the basis for the release of atomic energy. Rutherford was not very enthusiastic about the practicable possibility of atomic energy and dismissed the idea as 'moonshine.' The great discoveries of 1932 were not followed up; Rutherford did not see eye to eye with his colleagues in this matter, but Crowther thinks that Britain could not, in any case, have provided the great sums of money needed. The age of Big Science had begun, and Britain was to take a back seat for a few years.

When Rutherford died in 1937, there was a feeling that the Cavendish had concentrated unduly on nuclear physics and that other lines of research would have been more fruitful. Although nuclear physics declined in Cambridge, it flourished in other British universities during the post-war years and, by the late 1960s, was absorbing half of the Science Research Council's entire budget, leaving the other half to provide for the needs of chemistry, mathematics, astronomy, engineering, biology and the rest of physics.

Other lines of research were successfully developed in Cambridge. The new science of molecular biology began in the Cavendish, and it was there that the redoubtable Jim Watson and his collaborators discovered the double helix, which is the key to the genetic code and which has brought us closer to a scientific understanding of life itself. Another post-war development was radio astronomy—the study of feeble radio signals from outer space by techniques which have produced a new map of the universe and greatly advanced our ideas of how it all began.

Science
and
Society

Making Rules for the Road

The last man to have had the traffic situation under good control was King Sennacherib. Painted boards that he mounted along the processional way in Nineveh bore the peremptory inscription: 'Royal Road. Let no man lessen it.'

... he had the traffic situation under good control.

Today the scientists at the British Transport and Road Research Laboratory can tell us that a concentration of ten parked vehicles per mile reduces the effective width of a 26 foot carriageway by four feet, and diminishes its traffic-handling capacity at a speed of 15 mph by 275 passenger car units per hour. This is interesting enough, but what is to be done next? The King of Assyria had a simple answer. Anyone who parked a chariot within sight of a 'No Waiting' sign was promptly executed.

Gentler methods are favoured today, in the belief that the roads can be made to carry more traffic if only we can discover how to use them efficiently. Confronted by this assignment, the mathematician immediately realises that there is little to be gained by standing on the pavement and watching the cars go past. What he tries to do is to reduce the chaos of the carriageway to some sort of order by making a model which can be treated in a mathematical way.

The model is not a table top littered with tiny cars and plastic policemen; it is an abstract creation in which the research worker replaces the cars and drivers by concepts which can be more easily manipulated. Sometimes a stream of traffic is regarded as a liquid flowing through a pipe. An alternative method treats the cars as molecules in a gas. These two approaches give great scope for mathematical ingenuity and lead to results which are not as absurd as might be expected. Another effort, more easily understood, reduces the traffic situation to simple terms, considering a single stream of cars guided by intelligent drivers, undistracted by overtaking or intersections.

The first problem is to calculate the speed at which the maximum flow of traffic will be achieved. Obviously there is no flow if the road is empty, and there is no flow in a traffic jam. Somewhere between these two extremes the most efficient use of the road will be achieved. If the cars were all guided mechanically it would be possible to achieve a very high rate of flow. With human drivers a small but significant amount of time is needed for a change in the behaviour of one car to be translated into an appropriate response by the driver of the car behind. This means that if collisions are to be avoided a certain minimum space must be kept between cars.

Experiments show that a typical driver will take no notice of the car in front if it is more than 200 feet away. At shorter distances the driver reacts so that a difference in speed between his own car and the car in front is kept to a minimum. The rapidity of the following driver's reaction and the ferocity with which he uses his brakes or accelerator are individual characteristics which influence his judgment of a safe distance.

Measurement of these characteristics for typical cars and drivers allows the optimum speed to be calculated. In a few special circumstances— for example, in a tunnel—the predictions can be verified by direct measurement, and it is found that the agreement is reasonably good.

In a crowded single lane, a minor disturbance (produced, for example, when one driver slows down) is transmitted along the line, possibly causing some drivers to stop or to change speed rather violently. The disturbance dies away as the leading cars pick up speed again, but the combined effect of the delay is considerable.

This effect is of particular importance to the Port of New York Authority, which is responsible for the Holland and Lincoln tunnels carrying a great amount of traffic between New Jersey and Manhattan. Observing that the greatest rate of flow through the tunnel occurred on occasions when there were gaps of a few seconds duration here and there in the traffic stream, some of the Port Authority's mathematicians suggested that better conditions could be achieved by deliberately introducing breaks into the traffic stream. When this is done the shock wave still travels backward if the leading car changes speed, but its effect disappears as soon as a gap occurs in the traffic.

For the Holland tunnel, the maximum rate of flow by mathematical calculation is 1320 vehicles per hour. The average rate of flow over a period of 13 days, with no control measures, was 1176 cars per hour. On 12 other days the experimenters counted cars as they arrived during separate two-minute periods. If the total was less than 44 cars (corresponding to 1320 cars per hour), no action was taken, but if 44 cars arrived before the end of the two minutes, traffic was held up for the rest of the interval.

By this simple method flow rates of about 1300 vehicles per hour were achieved on several occasions, and the average for the 12-day test was 1248 cars per hour, representing a significant improvement.

So don't scowl when the traffic warden puts out his hand. If he has read the right books he will shorten your journey by holding you up.

Relation of Poetry to Science

Two cyclists are pedalling furiously along a country road. 'Where are we?' asks one. 'I don't know,' says the other, 'It's your turn to look up.' Many people would accept this story as a parable of contemporary science. When challenged, they would admit that the criticism applies less to science in general than to the individual scientists who lose contact with the outside world by undue concentration on their own particular problems.

Certainly the organisation of science today favours the specialist, but there are other factors maintaining the gap between the two cultures. When Dryden, Wren and Newton were active in the early days of the Royal Society, there were no barriers of language between science and art. In our own time the poet and the experimenter both work in realms of abstraction where communication with the uninitiated is difficult.

Can the gulf be bridged, as Bronowski advises in the terminal essay of Mr Eastwood's anthology†, by 'that one universal language which alone can unite art and science, and layman and scientist, in a common understanding?' Must more of the blame be laid on the scientist who, finding the task of popularising his speciality difficult or distasteful, persuades himself that it doesn't matter? Is the outsider simply too lazy to make the modest effort required to enjoy the pleasures of science?

Mr Eastwood illuminates these questions from the standpoint that 'poet and scientist are alike concerned with the ordering of experience and to both the imagination is all-important.' Since poetry is concerned with life, it must speak of science and technology. These views need not be disputed, but they have seldom been put into practice.

The anthologist casts his net widely, taking in agriculture and medicine to good effect. Admitting no distinction between poetry and verse, he includes much that satisfies the definitions made by Johnson ('metrical composition') and Newton ('a kind of ingenious nonsense'), but might not appeal to more fastidious judges. There is little to be said for Samuel Butler's *The Elephant in the Moon* (it turns out to be a mouse in the telescope), which rambles on for 14 pages. Another nursery joke by Oliver Wendell Holmes uses a fly in a stethoscope to encompass the downfall of a Boston physician, related with one touch of sick humour:

He shook his head;—there's grave disease—
I greatly fear you all must die;
A slight post-mortem, if you please,
Surviving friends would gratify.

† W Eastwood 1961 *A Book of Science Verse* (London: Macmillan).

Leaving aside a few curiosities like these, the anthology contains four kinds of poetry. First come Lucretius, Dante, Shakespeare, Milton and Pope on various aspects of the physical universe, with Chaucer and Ben Jonson on alchemy. The second and largest (though not the most attractive) group consists of poems or extracts with some topic in science or engineering for their inspiration. The lesser poets are much in evidence here. Some of their contributions are mere rhetoric, like Akenside's *Hymn to Science:*

> Science! thou fair effusive ray
> From the great source of mental day ....

Still less felicitous is the *Ode to Mr McAdam* by Keats' friend, John Reynolds:

> McAdam Hail!
> Hail, Roadian! Hail, Colossus! who dost stand
> Striding ten thousand turnpikes on the land!

Later verses applauding the

> Dispenser of coagulated good!
> Distributor of granite and of food

are even closer in style to McGonagall—but perhaps this particular ode was intended as a parody.

Poetry, said Wordsworth, is the impassioned spirit which is in the countenance of all science. Here he expresses a similar view, writing on *Steamboats, Viaducts and Railways* in 1833:

> Nor shall your presence, howsoe'er it mar
> The loveliness of Nature, prove a bar
> To the Mind's gaining that prophetic sense
> Of future change, that point of vision, whence
> May be discovered what in soul ye are.

'By angles and ratios harmonic he strove
Her curves and proportions all faultless to prove.'

By 1844, when the Kendal and Windermere railway was proposed, he had changed his tune:

> Is then no nook of English ground secure
> From rash assault?

Thirdly, Mr Eastwood offers samples of comic verse, including Macquorn Rankine (Professor of Civil Engineering at Glasgow University from 1855 to 1872) on *The Mathematician in Love:*

> A mathematician fell madly in love
> With a lady, young, handsome and charming:
> By angles and ratios harmonic he strove
> Her curves and proportions all faultless to prove.

He solves the equations successfully, but the lady runs off with a dragoon in the last verse. The compiler's researches might usefully have gone deeper, for quite a few scientists have expressed themselves in verse.

Whewell did so unconsciously in his *Elementary Treatises on Mechanics* published in 1819:

> And so no force, however great,
> can stretch a cord, however fine,
> into a horizontal line
> that shall be absolutely straight.

James Watt appears in E V Knox's *The Steam-givers*, a mischievous parody in the style of Alfred Noyes:

> He was a great mechanic, was James Watt,
> And ever as he toiled and murmured 'Steam'
> He sang some stave to wile the weary hours
> And break the page, some little stave like this—
> In Old Cathay, in far Cathay
> Before the West espied the gleam,
> Philosophers had found no way
> Of fruitfully condensing steam.
> With instruments that went by hand
> Their unenlightened path they trod;
> The Chinese did not understand
> The uses of the piston-rod†.

Quiller-Couch offers a page of Euclid in attractive form:

> The King sits in Dunfermline toun
> Drinking the blude-red wine;
> 'O wha will rear me an equilateral triangle
> Upon a given straight line?'‡

---

† From E V Knox 1973 *The Steam-givers* in *These Liberties* (London: Methuen).
‡ From A T Quiller-Couch 1906 *From a Cornish Window* (London: Cambridge University Press).

Sir Patrick Spens, summoned from Cambridge, provides the desired construction before falling foul of 'that envious knicht, Sir Hughie o' the Graeme.' Both are slain in a fight 'forenenst the Asses Brig.'

> But let us sing Long live the King!
> And his foes the Deil attend 'em
> For he has gotten his little triangle,
> Quod erat faciendum!

Finally, the anthology shows how contemporary poets have reacted to an environment dominated by science and technology. To judge from Mr Eastwood's sample, they have not reacted in any very obvious way. C Day Lewis describes the epic flight of Parer and McIntosh to Australia in 1920. Stephen Spender muses on *The Express* and *The Landscape near an Aerodrome*, and Patric Dickinson, looking at Jodrell Bank, writes

> Now
> We receive the blind codes
> Of spaces beyond the span
> Of our myths, and a long dead star
> May only echo how
> There are no loves or gods
> Men can invent to explain
> How lonely all men are†.

John Wain's electronic brain speaks sharply to its maker:

> You call me part of you. You lie,
>   I am
> Myself. Your motive building me,
>   was false.
> You wanted accuracy: figures,
>   charts.
> But accuracy is a limb of truth,
> A limb of truth, but not her holy
>   body.
> Must I now teach you that the
>   truth is one,
> Its accuracy of wholeness, centred
>   firm?‡

The lingua franca which is to unite art and science seems as far off as ever, but the quest can be instructive and entertaining.

---

† From Patric Dickinson 1960 *Jodrell Bank* in *The World I See* (London: Chatto and Windus).
‡ From John Wain 1956 *A Word Carved on a Sill* (London: Routledge and Kegan Paul).

Melvyn the Map

Geography, the poet observed, is about maps—in contrast to biography, which is about chaps. Professor Melvyn Howe, well known for his ingenious maps of disease mortality (which confirm the persistent belief that Bournemouth is a healthier place than Bootle), has enlarged the scope of geography with an interesting study† of the relation between disease and environment.

... persistent belief that Bournemouth is a healthier place than Bootle.

His concern is mainly with the plagues and pestilences of earlier days, but his narrative offers lessons for the present and the future. The last 100 years and, more especially, the last 30 years, have seen a spectacular revolution in public health through the control, the understanding and finally the virtual elimination from Britain of the killing diseases—bubonic plague, typhus, cholera, smallpox, tuberculosis and diphtheria—which raged unchecked during the preceding ten centuries. Life expectancy was probably about 30 years in Norman times, and little more at the beginning of the nineteenth century; today it is about 70 years.

Until recent times, most people died in childhood; in 1887 half of the deaths recorded in Glasgow were of children under ten years of age. Those who survived could expect epidemics at least once in every decade— and no effective treatment. Until late in the nineteenth century, writes Howe, 'medical men knew little more than had their Greek forbears about the actual causes of plague, fever and pestilential scourges.'

† G Melvyn Howe 1972 *Man, Environment and Disease* (London: David and Charles).

Although the importance of rats and insects as carriers of disease was not appreciated until after the plague and other such afflictions had been conquered, some inkling of the significance of environmental factors had been early apparent. Even in Tudor and Stuart times, the wealthier citizens fled to the cleaner surroundings of the countryside when infection threatened.

The physicians of Saxon England thought that diseases were caused by elfshot, or fairy arrows. A thousand years later, odours, vapours and miasmas were blamed—an idea which goes back to St Columba. Seeing a cloud rising from the sea, he predicted:

> This cloud will be very harmful to men and cattle ... it will pour down in the evening a pestilential rain which will cause grievous and festering ulcers.

Adamnan, Columba's biographer, claimed that monasteries were spared, but there is evidence that epidemics were spread by missionaries. The plague which began in 664 visited many isolated monasteries; the Synod of Hertford, meeting in 672, forbade monks to travel from one monastery to another 'after the Celtic fashion.'

The lesson was not remembered, and travellers continued to spread pestilence. The Black Death halted at the Scottish border in 1349 and smug comments were made about 'the foul death of the English,' until a raiding party brought the plague back into their own country, where it killed a third of the population. After the 1745 rebellion, English soldiers recalled from the Low Countries brought typhus to many parts of Scotland.

The plague disappeared from Britain after 1665 (no one knows why), and now survives only in the laboratory. In 1962 it killed a man at the Chemical Defence Experimental Establishment at Porton Down, recalling the beginning of biological warfare in 1346. In that year the plague first came to Europe: during the siege of Kaffa (a Christian city), the Infidels, smitten with plague acquired in Asia, used catapults to hurl their dead among the defenders.

What can we learn from the diseases of the past? Mainly that science, if it produces any answers at all, may be too late to influence events. Typhoid, cholera and smallpox were controlled before the organisms causing them had been seen under the microscope. Empirical solutions, including vaccination, better water supplies and sewage disposal, turned the tide. Tuberculosis, on the other hand, was declining rather slowly until effective drugs appeared 30 years ago.

History gives us no clues to help the attack on today's epidemics of cancer and heart disease—except, perhaps, to tell us that the answers are already there, if we could only see them.

Fall-out and Health

Peter King was a young chemist in 1948 at the Naval Research Laboratory in Washington, where he studied the chemical properties of paint.

One day, the Geiger counters used in some isotope experiments were counting faster than usual. This sort of thing happens quite often in laboratories using radioactive materials, and most people would think it hardly worth investigating.

Dr King suspected that the counters acted strangely after every heavy fall of rain. Most probably this hunch was not justified, but its consequences were remarkable. Samples of rainwater were examined and were proved to be slightly radioactive. This observation might not seem very significant, for any specimen of rain is measurably radioactive for an hour or two after it has fallen. The activity comes from short-lived isotopes of the radium and thorium families.

The Washington rain, however, remained radioactive for a good deal longer, and Dr King wondered whether it might have been contaminated by the American atomic bomb tests that had been carried out in the Pacific a few months earlier. Knowing that rainwater was collected assiduously in the Virgin Islands, and that the natives could generally remember when a particular cistern or barrel had been filled, he sent an assistant to gather samples corresponding to the time of the bomb tests and the period shortly after.

When these specimens turned out to be radioactive, and to contain isotopes of cerium and yttrium (which could only have come from a nuclear explosion), Dr King realised that he had found a cheap and simple method for identifying atomic weapon tests.

He was the first man to appreciate the importance of this idea and lost no time in putting it to practical use. The Russians were then believed to be well behind the United States in their development of atomic weapons. The full extent of their efforts in nuclear technology (much helped by scientific spies in Britain, Canada and the United States) was not known or acknowledged.

Dr King, with the scepticism of the true scientist, decided to find out for himself. He persuaded the US Navy to send him a sample of rainwater every month from Kodiak Island in the Gulf of Alaska. The Geiger counters showed no activity for quite a long time, but they began to click one day in September 1949. Chemical tests showed the presence of radioactive cerium again. A few days later, the news of Russia's first atomic explosion was released—not by the Russians, but by President Truman, acting on Dr King's findings.

These historic experiments were the first to measure fall-out, the radioactive debris produced in every nuclear explosion, which is subsequently distributed widely over the Earth's surface. In the early days, the work was of great scientific interest but seemed to have little practical importance. Megaton weapons were unknown, and no tests had been made with bombs of really large explosive power.

The situation changed dramatically with the detonation of the first hydrogen bomb at Bikini Atoll on 1 March 1954. In a fraction of a second the radioactive contamination of the atmosphere was raised to a previously unknown level. The hazard of radioactive strontium now had to be taken seriously. The isotope strontium-90 (which is chemically similar to calcium) descended over the Earth and was absorbed into food (particularly milk), cereals and other sources of dietary calcium.

It soon became evident that the experts had no reliable views on the possible danger of swallowing strontium-90. Discussions centred on the explosive power of the hydrogen bombs and on the contamination of food or human tissues, measured in Strontium Units. One SU is a concentration of one micromicrocurie of strontium-90 per gram of calcium.

An American study known under the name of 'Project Gabriel' advised in 1949 that the level of strontium-90 in the human body should not be allowed to rise above 10 000 SU. A later estimate predicted that this level would not be reached until bombs of more than 40 000 megatons explosive power had been detonated. (A megaton is a million tons; the hydrogen bomb is usually described by reference to the amount of TNT needed to match its explosive power.)

In 1953, Dr Hans Bethe declared that 8000 megatons would be needed to reach the danger limit for strontium-90 in the human body. An American working party in the same year advised that the maximum permissible concentration should be reduced to 1000 SU, and that bombs totalling 25 000 megatons could be released without reaching this limit.

Another expert witness, Dr W F Neuman, told an official enquiry in 1957 that 44 megatons a year would produce the same effect on a world-wide basis. Another committee recommended that fission products should not be released at an annual rate equivalent to more than ten megatons. The Director of the US Atomic Energy Commission's division of biology and medicine estimated in 1951 that the danger level for fission products was a total of 20 000 000 megatons. Experts often disagree, but seldom on such a monumental scale as this. Fortunately, the authorities responsible for the public safety in these matters took a cautious line in the face of so much contradictory advice.

There are reasonable grounds for believing that no serious harm has been done by the strontium-90 so far sprinkled over the Earth's surface, but it must be admitted that good luck has been an important factor in avoiding trouble. Monitoring programmes during the 1960s, when nuclear weapon testing was at its height, showed that the concentration of strontium-90 in human bone was much lower than in the vegetation from which our food is derived.

The difference arises firstly because the cow, for example, can reject a large part of the strontium which she picks up in the grass, delivering only 10 or 20% of it in her milk. A baby has the same power of discrimination to

Experts often disagree ...

a lesser degree, absorbing the calcium, but rejecting about half of the strontium-90. Older children and adults retain strontium to only about a quarter of the level that might be expected.

There is no very obvious reason why animals and human beings should be able to absorb calcium from their diet in preference to strontium. If the two elements were treated equally by the digestive system, the concentration of radioactive strontium in children's bones would have risen to 10 or 20 times the level actually found—well within the danger zone. If the body retained strontium in preference to calcium the outcome would have been even worse.

These considerations were not understood when the first hydrogen bombs were exploded. The strontium story is finished for the time being (the monitoring of bone samples on a meaningful scale ended in Britain in 1972), but its moral is plain. Tampering with the environment will often put the health of mankind at risk in a way that cannot be properly evaluated until the opportunity for corrective action has passed. We may not be so lucky next time.

Disco Deafness–the Hazard that Never Was

Just a few years ago, the learned (and even the not so learned) journals gave much attention to the risk of deafness among teenagers who frequented discotheques and musical salons of a similar kind. Earnest authorities (mostly middle-aged) offered magisterial warnings of acoustic perils lurking in the caverns of rock and pop.

When the talking stopped and the figures appeared, the evidence was so flimsy as to be almost worthless. In one group of about 450 students, 82% were rejected because their hearing was not up to much anyway. Among the remainder, the pop fans had hearing losses of 2–3 decibels, in comparison with controls who did not regularly visit discotheques.

A hearing loss of 2–3 decibels is insignificant, whatever the t-test shows. As Lord Rutherford used to say: 'If your experiment needs statistics, you should design a better experiment.'

It is, however, difficult to understand why exposure to sound levels of 90–100 decibels in a discotheque does no harm, while engineering workers exposed to the same sound level for the same time often show considerable hearing damage.

The explanation, recently provided by Dr P V Bruel, an eminent Danish engineer, is quite simple. He starts from the observation that a steady noise or sound offered to the ear for a short time (less than a fifth of a second) will appear less loud than it actually is; the shorter the sound, the weaker it appears.

The reason is that the brain samples a sound and does a little processing before deciding what to perceive. If a sound is too short, the sampling is incomplete. So a sudden peak of sound will be perceived at less than its true loudness, or may not register at all in the auditory cortex.

A similar effect is related to the aural reflex, which normally protects the inner ear from loud sounds. At sound levels above about 80 decibels, the auditory ossicles (which transmit vibrations from the eardrum to the inner ear) stiffen up and transmit less effectively— a reaction analogous to the response of the iris to bright light.

The aural reflex has a reaction time of a small fraction of a second, so that short peaks of sound will pass through the middle ear without invoking

While engineering workers ... show considerable hearing damage.

Disco deafness—the hazard that never was

the protective mechanism. Most sound-level meters are designed, naturally enough, to respond like ears, so they do not register the short-duration peaks.

The story now becomes clear. What damages the hearing is not the average sound level, but the peaks. Factory or shipyard noise, often impulsive in nature, has a lot of energy in peaks which are not recorded by a conventional sound-level meter (or perceived as sound in the brain), but can damage the hearing by destroying hair cells in the cochlea.

Disco noise has rather little energy in dangerous form, because the peaks are limited by the inadequate response of the amplifiers and loudspeakers. The same restriction applies to live performances, where most of the sound comes from amplifying equipment.

Dr Bruel concludes that any suspected risk to hearing should be assessed by measuring both the average sound level and the amount of sound energy in short-duration peaks. When this is done, disco sound is found to be relatively harmless, however offensive it may be to middle-aged ears.

Atomic Energy in War

> It might be as good as our present-day explosives, but it is unlikely to produce anything very much more dangerous.

Winston Churchill was writing (to Kingsley Wood, the Secretary of State for Air) about the atomic bomb, which, in August 1939, was already a subject of speculation in scientific laboratories and in the popular press. His judgment, based on advice from Professor Lindemann (afterwards Lord Cherwell), was consistent with the best available evidence.

A few months later Otto Frisch, writing in the *Annual Reports on the Progress of Chemistry* for 1939, concluded that the construction of a 'super-bomb' would be prohibitively expensive, if not completely impossible, and that its destructive power would be less than had been supposed.

Frisch (afterwards Jacksonian Professor of Natural Philosophy at Cambridge) might be claimed as the discoverer of nuclear fission, though the story is confused by many false clues and missed opportunities. Certainly he borrowed the word fission to describe the splitting of the uranium atom, and made a remarkably accurate calculation of the energy released in this process before assembling the simple apparatus needed to give experimental verification of his ideas.

After writing his article on fission, Frisch had second thoughts and soon became convinced that the bomb could indeed be made. With R E Peierls he wrote a three-page report setting out the theoretical basis of the atomic bomb, outlining the practical steps necessary to build it, and predicting its effects.

This report, one of the most significant documents in the long history of science, soon reached the Government's scientific advisors and caused a revival of interest in the uranium bomb. By April 1940, the Maud Committee had been established to advise on future developments. The committee included some eminent people, including Cockcroft and Blackett, but not the two men whose proposals had inspired it. Peierls had just been naturalised and Frisch was officially an enemy alien.

By the strict application of Government policy on refugee scientists, they were not suitable persons to have access to the secrets which they had themselves discovered. Fortunately, they kept their patience through this Gilbertian situation and were eventually made members of a technical sub-committee.

By the end of 1940, a large part of the work on the bomb was being done by aliens or recently naturalised refugees, including Simon, Rotblat, Bretscher, Halban and Kowarski. With some misgivings on the part of the Ministry of Labour, further reinforcement was provided by Kurti, Kuhn, Kemmer, Freundlich—and Fuchs (who was later convicted of espionage).

In August 1940, it was known that a chain reaction generated by slow neutrons would produce energy, but not at a rate suitable for a bomb.

With fast neutrons a chain reaction seemed possible in natural uranium, although the critical mass, below which nothing significant would happen, amounted to many tons. If the rare isotope uranium-235 could be separated from the more abundant uranium-238, the critical size would be only a few pounds and the bomb would in all probability be highly effective.

It was then recognised that the separation of the two isotopes of uranium was the crux of the matter. Simon and his colleagues in Oxford produced realistic plans for a gaseous diffusion plant, to cover 40 acres, with a capital cost of £4m, and running costs of £1·5m per year.

By the end of 1940, the bomb project had made astonishing progress. Uranium metal was being made by ICI, the possibility of plutonium as a nuclear explosive had been uncovered, and the building of the atomic weapons had come close to feasibility.

All this was achieved at very little expenditure of Government money—and in impressive secrecy. In February 1941, the Secretary of the Department of Scientific and Industrial Research (Appleton) knew little about the activities of the Maud Committee; and Lord Hankey, Chairman of the Cabinet's Scientific Advisory Committee, did not even know of its existence.

The first wartime contacts between British and American nuclear scientists were made towards the end of 1940, when Sir Henry Tizard and J D Cockcroft led a mission to Washington. American work, which was freely described to the visitors, was found to be well behind the British effort. Nevertheless, misgivings arose about the desirability of continuing the project to its ultimate conclusion in Britain. Cost was an important factor, but manpower problems, shortages of materials, and the danger of air attack proved to be decisive factors, and Britain handed over the lead.

There were, too, some moral doubts. One of these was contained in a letter written by Dr (later Sir) Charles Darwin, then Director of the British Central Scientific Office in Washington, in 1941 asking that if the bomb

American work . . . was found to be well behind the British effort.

could be made, would the US President and British Prime Minister be willing to sanction the total destruction of Berlin and the country round it when, if ever, they were told it could be accomplished in a single blow.

Another anxiety was reflected in a note of a talk between Lord Halifax, then British Ambassador to the United States, and Mr Henry Stimson, US Secretary for War, when Lord Halifax asked of the possibility that before the bomb was dropped, 48 hours' warning should be given to the Japanese, offering the alternative of unconditional surrender.

For a while in 1941, it seemed that British and American scientists might work in a joint programme. Post-war political problems were already being recognised. Lord Cherwell wanted the uranium separation plant to be built in England. Whoever possessed such a plant, he advised, would be able to dictate terms to the rest of the world. He did not want the United States to play this role, and Lord Hankey agreed.

Further consideration re-affirmed the policy of close collaboration with Canada and the United States, with the major manufacturing facilities to be built in North America. In the six months after Pearl Harbor, the American effort grew rapidly. Friendly and effective collaboration continued during most of 1942, but broke down towards the end of that year, when the military took command of the project.

From then on, the Americans expected to receive British information but, over most of the project, to give nothing in exchange. 'Quite intolerable,' wrote Sir John Anderson to the Prime Minister, 'You may wish to ask President Roosevelt to go into the matter without delay.' He did not know that Roosevelt had himself approved the restriction of collaboration.

Later the British and American viewpoints were reconciled to some extent in the Quebec Agreement of 1943, which called for 'full and effective' collaboration. It soon became clear that Britain was to be the junior partner. British scientists joined the teams at Berkeley, California, and at Los Alamos, New Mexico, but the major effort came from the United States.

American feeling was quite strongly against British participation in work with an actual or potential industrial background, but less restrictive in regard to purely scientific activity. For this reason the British knew a great deal more about the Los Alamos project, from which the first uranium and plutonium bombs were to emerge, but not so much about the uranium separation work and the nuclear reactors built in the United States.

After 1945, however, Britain took a leading part in the exploitation of atomic energy, including the development of a substantial nuclear power programme (using both fast and thermal reactors), and a thriving business in radioactive isotopes for industrial and medical purposes. But early estimates of the export market for nuclear power stations were not fulfilled, and Britain's pioneering efforts did not gain the rewards that they deserved.

Science and Society

What makes the scientist tick? Is he a hero of the intellect, or merely a reflection of the prevailing cultural and technological background? Science, strictly defined, is certainly an intellectual activity. It deals with abstract ideas and concepts, using them to make models which—by recourse to experiment—can be tested against reality. When the scientist steps out of his study and into the laboratory he is seeking the help of technology. Science, in today's jargon, is software; technology is hardware.

Technology is inevitably linked with society since, using materials and crafts, it involves many people. Science might be regarded as a self-contained activity, independent of the contemporary social structure, but this view is now of very limited validity. Some of the more remote reaches of mathematics are closed philosophical domains, concerned merely with the properties of numbers, but almost every other branch of science has come to terms with society.

The study—and indeed the recognition—of this process has been pioneered by Robert K Merton of Columbia University. *The Sociology of Science*† is a welcome collection of 22 of Merton's essays, originally published at various times over the past 40 years.

His first major work was *Science, Technology and Society in Seventeenth-Century England*. The choice of period is significant, for it marks the beginning of modern science and, in a related but less obvious way, the beginning of modern history.

The seventeenth century was the time when (led by Isaac Newton) mathematics took charge, when philosophical speculation gave way to calculation and measurement, and when the study of nature changed from anecdote to analysis. There were mathematicians at work before Newton (notably Galileo and Descartes), but their efforts to reconcile the mechanical tradition (seeking a cause for every effect) with the geometrical description of nature ultimately derived from Pythagoras, had reached an impasse. Newton bridged the gap with the idea of action at a distance—for example, in gravitation. This innovation was resisted because it smacked of the occult, but its successes were so impressive that the opposition melted away.

Merton is concerned less with these trends of thought than with the social factors influencing scientific progress. The growth of science and technology were, he claims, encouraged by the prevailing Protestant ethic and, in particular, by Puritanism. There were three reasons. Firstly, the Puritan was haunted by the doctrine of predestination. Only the elect would be saved; but how could a man know whether he was of the elect? The answer lay in good works, which provided outward evidence of inward grace. To a Catholic, good works often meant meditation in a monastery; to

---

† Robert K Merton 1973 *The Sociology of Science* (Chicago: University of Chicago Press).

a Protestant, the words denoted useful (and often profitable) activity in the world. The Protestant sects which did not believe in predestination reached the same conclusion, regarding diligence in good works as a way of reaching a state of grace, rather than of demonstrating what was already there.

The answer lay in good works.

The Puritan had to be busy about his work, since idleness conduced to sinful thoughts, play-going and flesh-pleasing. How much better to occupy the mind (declared Thomas Spratt) in the pursuit of science which 'can never be finished by the perpetual labours of one man .... It makes us serviceable to the world.'

Secondly, the Puritans believed that the glory of God could be best promoted by the close study of His works. To Robert Boyle, as to Francis Bacon earlier in the seventeenth century, and to many of their contemporaries, the practice of science was essentially a religious activity.

Thirdly, it was claimed (and, before long, demonstrated) that man's earthly lot could be improved by inventions allowing the control or exploitation of nature. The Charter (still in force) of the Royal Society, the world's greatest scientific academy, ordains that the efforts of the Fellows shall be directed 'to further promoting by the authority of experiments and the sciences of natural things and of useful arts, to the glory of God the Creator and the advantage of the human race.'

Merton has examined the social origins of the founders and early Fellows of the Royal Society. Of those whose religious leanings are known, the majority were Puritans. The same influence helped to establish science in America. The founders of Harvard College were imbued with the spirit of Calvinism; there and elsewhere in New England the Puritan divines encouraged the growth of science.

Although inspired by religious beliefs, the new science was quickly applied, in England, to mundane problems. The seventeenth century was a

time of war and revolution, and a time of rapid change in military technology, as the sword and the pike gave way to firearms, and the cannon became
an important weapon. The Royal Society, as Merton shows in a detailed
analysis, was much concerned with the improvement of gunpowder, the
flight of bullets and cannon balls, the recoil of guns, and the casting of
metals. This work led to scientific studies (of no direct military importance)
on the compression and expansion of gases, the strength and elasticity of
metals, and the nature of explosions—early examples of the way in which
science springs from technology.

Social and economic needs guided the development of science and
technology during the eighteenth century. The shortage of wood led to the
development of the coal industry. When the lack of water and difficulties of
terrain curtailed the growth of the canal system, merchants and coal-owners
created railways. The expansion of coal-mining created drainage problems
which horse-driven pumps could not solve. The challenge was first met by
Newcomen's steam engine, and later by Watt's improvements which were so
important in the Industrial Revolution.

Merton does not dwell on the eighteenth and nineteenth centuries,
but takes a close look at twentieth-century science. The method that he uses
is to examine the ways in which scientists interact with one another and with
society. In other words, he tries to construct a sociology of science. This has
been his life's work; he has done it so well that, until recently, few scholars
joined him in the study and development of the new discipline.

Approaching the problem as a behavioural scientist, he asks how
scientists mark out their territories, how they communicate with one
another, and how they establish pecking orders. His main theme here relates
to priority of discovery. A rough and ready way of rewarding scientific effort
is to give honours or prizes to those who first discover (or, to be more exact,
those who first announce), for example, a new atomic particle, a new
scientific law, or a new explanation for some natural phenomenon. Since the
history of science is usually written as a sequence of discoveries, anxiety to
be first is not surprising.

Merton shows that the simple concept of the scientist as hero is not
valid; too many discoveries are made independently (and sometimes almost
simultaneously) by two or more people. Does this mean that discoveries are
inevitable and are determined by the environment rather than by the
individual? Is the work of all but the first discoverer wasted? Merton
resolves this conflict by showing that the greatest men of science have been
involved in multiple discoveries. Lord Kelvin himself noted 32 occasions on
which he eventually found that his independent discoveries had previously
been made by others. The true score must have been even greater, since we
do not yet know how many of Kelvin's discoveries were afterwards made
independently by others.

Galileo, Newton, Faraday, Maxwell and many other scientists of
genius had the same experience as Kelvin. But that does not mean that they
were superfluous and that things would have turned out much the same

without them. A scientist of genius is involved in multiple discoveries simply because he discovers so much. Many of Kelvin's firsts remained firsts: his 32 duplicates involved 30 other scientists, some of whom achieved nothing else of importance. So (says Merton) a great scientist, even though some of his discoveries are anticipated or re-discovered, is functionally equivalent to many lesser men.

The scientist still on the way up, not knowing whether history will declare him a genius, usually tries hard to establish priority for any original work. This can be done by publication in a scientific journal, in the programme of a conference, or in a report circulated to selected laboratories, including the laboratories of any competitors who, if they are gentlemen, will thus be disarmed. There are other ways of beating the clock, however. The simplest is to select—and publish—only the experimental observations which fit the theory under test. If the theory eventually proves to be valid (possibly after further long and difficult research by other people), the priority of the first announcement is confirmed; if later work does not support the claim, the matter may be soon forgotten; the scientific path abounds in artefacts and dead ends.

The most eminent exponent of this kind of short cut was Gregor Mendel, the father of genetics. The results of his plant-crossing experiments were too good to be true. The deception was not pointed out until long after Mendel's death. Oddly enough, Mendel was not concerned about priority; his researches had no effect on his contemporaries. He died (in 1882) a respected but obscure amateur, and the importance of his work was not recognised until 1900, after his theory had been re-discovered simultaneously in Germany, Austria and Holland. Another technique is to alter the theory to fit the experimental results. Newton was an expert in fraud of this kind; but since he was usually right in other ways, his successors have cheerfully ignored (or whitewashed) his trickery.

Earth, Sun and Stars

Big Magnet

An ingenious inventor wrote to the Admiralty in 1940 suggesting a simple method of winning the war. His idea was to charter telegraph cables to form a continuous loop as near as possible to the equator. The two ends were to be joined to a battery; when this was done the resulting electric current would reverse the Earth's natural magnetism, thereby detonating simultaneously every magnetic mine in the seas.

The detailed calculations accompanying this bold proposal were (if legend is to be believed) approved by several braided mariners before being subjected to scientific scrutiny. It was then discovered that the inventor had become confused between two systems of electrical units, and that the current required to fulfil his proposal was so enormous that no cable could possibly carry it.

Less spectacular methods of combating the magnetic mine were adopted with considerable success, but the reversal of the Earth's magnetic field does not look quite so absurd now as it did then. In recent years, evidence has been accumulating that the magnetic poles of the Earth have changed places many times in the relatively short time of a few million years.

The Earth's magnetism was discovered by William Gilbert, an eminent practitioner of alchemy and medicine, who was physician to Queen Elizabeth I from 1601 to 1603. He also served King James I even more briefly in the same capacity, and died of the plague in November 1603.

In seeking a scientific explanation for the behaviour of the compass needle, which was well known to navigators at that time, he observed that a magnetised needle, placed on a vertical pivot, will point roughly towards the north. If, however, the needle is held on a horizontal pivot, so that it can swing in a vertical plane, it will almost always come to rest pointing obliquely downwards into the Earth.

Near the equator, the inclination of the needle is quite small, and in northerly latitudes it may point almost vertically downwards. The source of this attraction was thought by some scholars to be in the sky, but Gilbert found the true explanation with the help of a simple model consisting of merely a ball of lodestone (a magnetic rock) and a compass needle. He found that the needle, when brought to various parts of the lodestone ball, pointed downwards in much the same way as if it had been moved over the surface of the Earth. He decided that the Earth was itself a magnet, a conclusion which has been accepted ever since, although it is still not fully explained.

The mariner's compass needle does not point to the geographical north, but in a slightly different direction, known as the magnetic north. The difference between these two directions usually amounts to several degrees, and changes quite perceptibly over a period of a few years.

Variations in the Earth's magnetic properties over a longer period may be uncovered by the study of rocks. Much of the rock now lying about the surface of the globe is the result of volcanic action in ancient times. In the process of cooling and solidification, rocks which contain iron (as most of them do) take on a permanent magnetism corresponding to the direction and intensity of the Earth's magnetic field at the appropriate time. This magnetism has survived unchanged for millions of years, and can now be measured in the laboratory.

The original magnetism is augmented by the effects of lightning, which may be removed in the same way as a watchmaker demagnetises a watch. The rock specimen is put inside a coil, through which an alternating current is passed. The rock is magnetised first in one direction and then the other, going through 50 cycles every second, while the current is gradually reduced to zero. The magnetisation of the specimen goes through ever-decreasing cycles until it too has vanished. Fortunately, this cleaning trick removes the unwanted magnetism of more recent origin, but leaves the original magnetic history of the rock unchanged.

In the early years of the twentieth century, when the measurement of rock magnetism was first studied seriously, it was generally found that rocks composed of recent volcanic lava, as well as samples of baked clay from archaeological sites, were magnetised roughly in the same direction as the Earth's present magnetic field at the sites where the samples were taken.

When Bernard Brunhes discovered a lump of rock (in 1906) which was magnetised in the opposite direction to that in the surrounding rocks, some people found the evidence hard to believe, while others left it alone as a curiosity.

Fresh interest arose in the 1950s, by which time more information and better techniques were available. Louis Néel, a French physicist, found from theoretical considerations that certain minerals would, when cooled from a high temperature, become magnetised in the direction opposite to that of the surrounding magnetic field. Almost immediately a group of Japanese scientists found a piece of volcanic rock which showed exactly the behaviour that Néel had predicted.

For some time it was not clear whether the Earth's magnetic field had reversed itself from time to time in past ages, or whether the samples of rock with reversed magnetism were simply materials having the chemical and physical structure which, as Néel predicted, would lead them to take on a magnetic imprint opposite to the normal direction.

This problem is not yet completely resolved, but it now seems probable that the north and south magnetic poles of the Earth have indeed been reversed several times during the last four million years. From the systematic examination of samples of different ages, it seems that all rocks up to about 700 000 years old are magnetised in a direction corresponding to the present magnetic field of the Earth. The same is true for most rocks between 2·5 and 3·5 million years old. In the age range between 700 000 and 2·5 million years, most of the rocks are magnetised in the opposite direction, with two groups (one with an age of about a million years and the other close to two million years) having a present-day magnetic signature.

The only plausible explanation for these findings is that the Earth's magnetic field has reversed itself, since there is no reason why rocks with the unusual chemical structure predicted by Néel should appear simultaneously in various parts of the globe, and then disappear again.

No one is able to explain why the Earth's magnetic field should change direction, or, for that matter, to explain why the Earth has a magnetic field anyway. The Earth's core is generally believed to contain a lot of molten metal. It is possible that slow churning might generate currents sufficient to account for the presence of the magnetic field, and even for the small variations which occur from year to year, but not for complete reversal.

Is it possible that a really violent shake-up might upset the liquid enough to generate fresh currents and to reverse the magnetic field? The Earth is continually bombarded by meteorites, some of which have left large craters, but fortunately these have all been in remote places.

Another group of missiles is the tektites—lumps of glassy rock first found in Czechoslovakia, but which have since been discovered in many other places. In composition, the tektites do not resemble any known rocks or volcanic material. No one has ever seen them falling, and several theories have been proposed to account for them. Some people think them to be meteorites, while others suggest that they have come from the Moon.

Another group of missiles is the tektites ...

A large bank of tektites, all of about the same age, has been found in Australia and the East Indies. Probably they all arrived at about the same time. In 1967, a group of American geologists reported that a large number of tektites had been found on the ocean bed in the same general area. The total weight of the tektite shower is now estimated to have been 150 million tons.

The impact of this amount of rock could, it is suggested, have disturbed the Earth's core severely enough to have reversed the magnetic poles. The theory is supported (or perhaps inspired) by the fact that many of the newly discovered tektites come from places close to rocks showing reversed magnetism.

If the Earth's magnetic poles change places again we shall know all about it. The magnetic compass is no longer necessary for serious navigation, but a reversal of the Earth's magnetic field would remove (temporarily at least) some of the protection which we now enjoy from cosmic rays. These rays (actually sub-atomic particles of great energy) are largely deflected by the magnetic field. It is in this way that we avoid the unpleasant radiation doses that would otherwise be incurred.

There is a little evidence that substantial evolutionary changes have taken place at times when the Earth's magnetic field was undergoing reversal in the past; radiation might have been responsible. Magnetic fall-out is one of the many mysteries that lie under our feet. Until more research is done, we shall not know whether it is likely to be a serious hazard for future generations.

Festival Day for Vulcan

There are not many blacksmiths left. Those who have survived the decline of the horse and cart probably go about their business as usual on 25 August, because no one told them that this date was long ago set aside for Vulcanalia, the festival ordained by the Romans to honour the founder of the craft.

Vulcan, the God of Fire, had his forge under Mount Etna in Sicily, with branch establishments at many other places, each conveniently provided with a volcano by way of ventilation. Ancient peoples used colourful legends to decorate what they did not understand. Today we wrap our ignorance in the trappings of science, and sometimes manage to produce quite a plausible story.

Vulcan, the God of Fire, had his forge under Mount Etna ...

Some scientific problems are difficult because they deal with such things as atoms, which no one is ever going to see. Volcanoes, on the other hand, are obvious and numerous. There are about 500 active (or potentially active) examples and thousands of extinct specimens, among them Arthur's Seat in Edinburgh, with its subsidiary features including the Calton Hill and the Castle Rock.

Live volcanoes discharge products of three kinds. Firstly, there is lava or molten rock. Volcanic action was much more vigorous in ancient times than it is today; consequently, large areas of the Earth's outer layer (including much of Scotland) have been deposited in the form of lava. Secondly, an erupting volcano produces huge amounts of steam, carbon dioxide, nitrogen and foul-smelling vapours. Thirdly, as these volatile materials force their way to the surface they break up the overlying rocks and scatter them for many miles.

In studying remnants of ancient volcanic incidents, geologists find lavas of rather similar composition (with silicon and magnesium among the most abundant elements) in many different parts of the world. These observations suggest that volcanic matter has a common origin, somewhere deep inside the Earth.

It is quite difficult to form any clear idea as to the internal structure and composition of the Earth, since most of it is beyond the reach of direct observation. We do, however, have a rough picture, based on measurements of the speed and direction of earthquake waves. Moving out from the centre of the Earth, the first 2000 miles make up the core, which is very dense and is thought to be composed mainly of iron and nickel.

The temperature of the core is more than 3000 °C, which is enough to melt any metal, but the pressure down there is so great that the material may still be in a solid condition. The next 1800 miles or so constitute the mantle, in which magnesium and iron predominate. The crust, which forms the outermost layer, may be 20 or 30 miles thick, although in some places it has a depth of only a few miles.

The composition of volcanic lava suggests that it comes from the upper layers of the mantle or perhaps the lower layers of the crust. The processes which bring it to the surface are still a matter for argument, but the story seems to be something like this.

A pocket of molten rock and gas (known to the geologist as magma) is formed in the lower part of the Earth's crust as the result of some disturbance which upsets the equilibrium of the mantle. This disturbance will usually be detected as an earthquake, though the shock waves may not be of any great violence when they reach the measuring instruments at the Earth's surface. Even though the temperature in the mantle is above the normal melting point of the rocks, the pressure is so great that no lique-faction occurs.

If the pressure is released, because of cracking or some other disruption, the rock will revert to the liquid state and will form a pocket of magma as soon as it escapes from the mantle. This pool of volcanic material may remain dormant until some weakening of the overlying crust allows it to escape. Alternatively, it may blow itself up to the surface by a simple process. As the magma cools and crystallises, some of the dissolved gases escape and build up a considerable pressure, which may be enough to break through the overlying layer of crust.

Volcanoes have been responsible for many disasters and have probably killed a quarter of a million people during the last 500 years. There are a few beneficial effects, including the very fertile soil on the slopes of Mount Etna, and the abundant supply of steam which now provides central heating for about a quarter of the houses in Iceland. Live volcanoes are better avoided, although dead ones can tell us a lot about the history of the Earth.

Sundials Tell More Than the Time

Ahaz, who was King of Judah some 2700 years ago, built a fine sundial. When his son Hezekiah was sick unto death, the Lord reassured him by causing His servant Isaiah to perform a miracle, moving the shadow on the sundial back by ten degrees, as we may read in the Second Book of Kings (20:11), and again in Isaiah (38:8).

This achievement has never been satisfactorily explained (if it had been, we could hardly count it as a miracle), but Sir Alan Herbert has offered some characteristically mischievous ideas on the subject in *Sundials Old and New or Fun with the Sun*†, a book of instructions for the do-it-yourself skiaphilist, with digressions on the beastliness of British Summer Time, the interior decoration of taverns, and the hardship inflicted on authors by public lending libraries.

Ahaz was probably one of the first men to make a sundial, taking advantage of a good climate and a general respect for technology. The dial was not very accurate by modern standards, but an error of a few minutes did not upset anyone in those days; indeed, it was customary in as late as the sixteenth century to use a sundial to check the performance of clocks and watches, which were even less exact.

The sundial is basically a very simple device, but there are several reasons why it cannot be a very good time-keeper, however skilfully it is made. It would in any case be wrong to regard the sundial merely as a rather poor sort of clock. It is much more than that; every sundial is a little replica of the Earth, offering a lot of information about the relationship between our planet and the Sun.

This achievement has never been satisfactorily explained . . .

† Alan Herbert 1967 *Sundials Old and New or Fun with the Sun* (London: Methuen).

This point can best be appreciated by making (or imagining) a rather special kind of sundial, consisting merely of a globe. It is best to take a specimen of reasonable size (at least six inches in diameter) and, if its mounting does not allow free movement in all directions, to unship it and support it on an empty can or dish of suitable dimensions.

Now adjust the globe so that its axis (the line joining the north and south poles) lies in the north–south plane, which can be identified from a large-scale map or with the help of a compass, remembering to make the necessary correction for the difference between the magnetic north and the geographical north. The globe should next be rotated about its axis until the line of longitude through your own position lies in the same north–south plane. (Lines of longitude are circles passing through the north and south poles.) Finally, rotate the globe about an east–west axis until the point representing your own position on the map comes to the top. When these adjustments have been made, a line from the centre of the globe to the highest point in the sky (known as the zenith) will pass through the point representing your own position on the map.

The globe should of course be placed out of doors. When the sun is shining, half of the globe will be illuminated—corresponding to the half of the Earth enjoying daylight at that time—and the other half will be in shadow. As the sun moves through the sky—and through the seasons of the year—the bright and dark halves of the little globe will change accordingly.

To tell the time, stick a couple of pins into the globe, one at the north pole and one at the south pole, each pointing along the direction of the axis. The globe may already be marked with lines of longitude; if not, draw a circle around each of the poles and divide its circumference into 24 equal parts. The position of the shadow of the pin will then indicate the time. The pin at the north pole will be illuminated during the summer, but will be in shadow during the winter; the shadow of the pin at the south pole then shows the time.

A sundial of the familiar kind is a simplified version of the globe that has just been described, consisting of the pin and not much more. The rod used to produce a shadow (sometimes called the style or gnomon) must, like the pin in the globe, point in a direction parallel to the Earth's axis. This means that the style must be vertical for a sundial at the north pole, and nearly horizontal in the tropics. At any other place the style should be inclined to the horizontal at an angle equal to the local latitude.

A sundial, naturally enough, keeps time by the Sun in an old-fashioned way, which seldom matches the clock. Today it is only a curiosity or, at best, a slightly eccentric hobby, but it can still teach us useful lessons in astronomy in an entertaining way.

(Herbert's explanation of the miracle at the beginning of this story is that Isaiah surreptitiously altered the inclination of the style; not very convincing, but no one seems to have thought of a better way of explaining the inexplicable.)

Back to the Beginning

The book of Genesis describes the creation as a process completed within a week: plants on one day, birds and fishes on another, and man (and woman) as the climax. Until about a century ago scientists accepted this narrative, and bent their minds to more pressing problems such as the steam engine and the electric current.

It might be thought that there are plenty of urgent problems to keep them busy today, without trying to rewrite the Bible. However, speculation about the origin of the Sun and the planets does help in understanding (and perhaps exploiting) the chemical composition of the Earth that we have inherited. There is no need to discourage the cosmologists, nor to switch off the computers which they use with the patience and enthusiasm that their predecessors applied to the telescope.

The Earth and the other planets are the cinders of a fire that has long been dead. It is difficult to trace the origin and course of the conflagration, because the nuclear fuels which supported it can burn in curious ways.

Lacking the imagination of the men who wrote the Old Testament, scientists do not go right to the beginning, but start with a large mass of hydrogen scattered through space; no one tries to guess where the hydrogen came from. What happens next is that hydrogen atoms come closer together because of gravitational attraction. The energy which they lose in this way is converted into heat, and since there is not much opportunity for this heat to escape (except by radiation), the temperature of the primitive star gradually rises.

Scientists do not go right back to the beginning ...

When the temperature reaches a few million degrees the hydrogen starts to burn. Four atoms of hydrogen combine to produce one of helium, releasing a good deal of energy, in a process which will go on for many millions of years. As the hydrogen is used up, the remaining atoms are drawn closer together by gravitation, and a further supply of energy is liberated. When the temperature reaches about 100 million degrees, helium atoms combine, three at a time, to give carbon. Further helium atoms are added on, producing oxygen and a number of other elements up to calcium.

By this time, the temperature has risen to about 1000 million degrees, and a number of other processes have begun. Many of the remaining chemical elements are formed by the successive addition of neutrons—a process used today on a more modest scale in the production of radioactive isotopes inside a nuclear reactor.

When the temperature of the star reaches about 5000 million degrees, the nuclear burning becomes very violent and generates a great number of unstable isotopes which blow up in a variety of ways until eventually all of the matter in the star is converted to an isotope of iron, which is the most stable of all atoms. For a small star, such as our own Sun, this process takes place quite slowly. When all of the hydrogen has been consumed (only a little of it has gone so far), the Sun will cool into a mere lump of iron.

For a big star, the corresponding nuclear reactions may be so vigorous that the core changes to iron in only a few minutes. This upheaval causes a violent explosion and the star is then known as a supernova. The products of the explosion are blown into space and may help in the formation of new stars.

Three supernovae have been observed in the Milky Way, which is within easy reach of our own telescopes. One appeared in 1054 and was recorded by Chinese astronomers, the second in 1572, and the third in 1604. A good many other supernovae have been observed in distant galaxies in recent years.

The explosions are so remote that they are not very spectacular as viewed from the Earth, but it has been noticed that the light from such an event falls to half of its intensity in 55 days, to a quarter in 110 days, and so on. This form of decay is characteristic of radioactive materials, and the half-life of 55 days corresponds to that of the element californium-254. This is an isotope which can be made in the laboratory in small amounts, but does not occur naturally in the Earth's crust. It is, however, believed to be the final product of the neutron capture processes occurring in a very hot star. Its presence in the supernovae suggests that star formation is still going on in many parts of the universe.

Another clue of the same kind comes from the presence of technetium-99, which has been revealed in many stars by examination (with a spectroscope) of the light that they emit. This isotope does not occur naturally on Earth, but its presence in a star can be accounted for by the nuclear burning processes already described.

The processes of star formation are, of course, still going on in the Sun even though they have long since ceased in the Earth. Broadly speaking, however, the Earth and the Sun show many similarities in chemical composition; these similarities are shared by meteorites, which were originally formed at the same time as the Sun and the primordial solar nebula, and can, whenever they reach the Earth in conveniently sized chunks, be subjected to chemical analysis. The Earth has, of course, lost a good many volatile elements and gases, but the relative abundances of the various chemical elements found in the crust and the oceans can be explained reasonably well in terms of the nuclear burning processes already described.

The time scale of scientific creation extends for a lot longer than a week, and the intermediate stages are rather complicated, but the scientist's narrative still begins at the moment when the Lord said: 'Let there be light.'

A Hazard as Old as the Earth

The atmosphere is a battlefield between the angels and devils ... the aspiring steeples around which cluster the low dwellings of men are to be likened, when the bells in them are ringing, to the hen spreading its protecting wings over its chickens; for the tones of the consecrated metal repel the demons and arrest storms and lightning.

The protective strategy recommended by Thomas Aquinas survived into the twentieth century in parts of Austria, illustrating that fiction is stranger than truth. While it was prevalent—until about 200 years ago—churches were destroyed and bell-ringers were electrocuted in thousands.

Safer methods of protection had to await better understanding of the nature of lightning. Anaximander wrote of 'the wind, enclosed in a thick cloud, which, by reason of its lightness, breaketh forth violently,' and other speculations invoked the spontaneous ignition of vapours of uncertain origin. When static electricity became popular in gentlemen's laboratories and ladies' salons during the eighteenth century, some experimenters noted the similarity between the sparks from their influence machines and the lightning flashes which occasionally brightened the sky.

Benjamin Franklin put this idea to the test and, with the help of a kite, showed that lightning is an electrical discharge from a cloud.

Usually the lightning passes between a cloud and the ground, but lightning flashes can also occur between two clouds. The base of a cloud is negatively charged. As it moves an electrical shadow follows it at ground

... fiction is stranger than truth.

The atmosphere is a battlefield ...

level, in the form of a corresponding positive charge. If the cloud passes over a high building the positive charge naturally rises to the top of the building and the insulation of the intervening air may break down.

The lightning flash, like any other electric current, always finds the easiest path—in this case the shortest—to earth, and that is why church steeples, trees, tents, and golfers with upraised clubs are particularly vulnerable.

Franklin was the first to suggest a rational method of protection against lightning. In as early as 1750 he proposed

> ... to fix on the highest parts of the edifices upright rods of iron, made sharp as a needle and gilt to prevent rusting, and from the foot of these rods a wire down the outside of the building into the ground .... Would not these pointed rods probably draw the electrical fire silently out of a cloud before it came nigh enough to strike?

The lightning conductor is very effective, though it does not work exactly as Franklin suggested. What actually happens is that the insulation of the air collapses just under the cloud and an electrical streamer starts on its way to earth. The streamer brings the electric charge from the cloud nearer to the ground creating such a powerful electrical tension that an upward streamer starts from the ground (or, more easily, from some high object) to meet the downward spark, and to provide a path for a heavy current.

If the rising electrical streamer comes from a lightning conductor, the eventual flash will be passed harmlessly to earth. If an unprotected building is struck, however, the sudden heating produced by the current can split stones or shatter wooden structures.

Ships' masts were often destroyed; hundreds of warships were lost—some in as late as 1843—while their Lordships debated the efficacy of Franklin's rods. The admirals were never convinced, but the advent of iron ships solved their problems.

At ground level, the safest places to shelter are in a ditch (preferably a damp one) or in a car with a metal body and roof; the lightning current will be conducted to earth by the tyres. The legend that the inside of a moving car becomes filled with static electricity which can be discharged by a dangling chain is nonsense—but that is another story.

Superstition is not confined to unlearned bell-ringers and car drivers. It is sometimes claimed that the efficacy of a lightning rod can be vastly increased by sticking a little radium on the end. The idea is that the radium ionises the surrounding air and facilitates the passage of a current. Unfortunately, theory and practice do not coincide; if they did, every hospital x-ray department would be in ruins.

Fortunately, a modern steel-framed building does not need external protection. If the structural steel is properly bonded together and continued into the ground, the worst that can happen is the loss of a chunk of masonry, after which the current finds its way to earth through the girders. A metal roof, if connected to the ground (for example, by rainwater pipes), acts as an effective lightning conductor. For this reason many old churches have survived, even though they have often been struck by lightning.

A rooftop television aerial is often used as a path for lightning and is sometimes responsible for damage; but the risk is small in relation to the hazard of mental and spiritual corruption from watching too much television.

Cultivating the Sea

'What I tell you three times is true,' the Bellman said. Consequently, no one doubts that two-thirds of mankind suffer from malnutrition.

Lord Boyd Orr's original claim, made in 1950, was based on an arithmetical error, but the Food and Agriculture Organisation, from which he had just retired, was still (as *The Economist* described it) 'a permanent institution devoted to proving there is not enough food in the world.' The FAO later modified the gloomy picture by pronouncing that half of the world was malnourished.

The evidence supporting these assertions was hard to find, and looked even less convincing when eventually the FAO disclosed that their calculations took something equivalent to the average British or French diet as the borderline of malnutrition. Even though millions of people live in reasonable health on diets that would not be approved by a committee of well fed experts, there is certainly a great deal of malnutrition.

It is commonly believed that the sea will rescue us from this predicament and that the nourishing soup which covers most of the globe can, with a little encouragement, produce enough food for all. At first glance there is certainly room for improvement. Although the sea covers nearly three-quarters of the globe, it produces only 2% of our food. Why is the productivity of the oceans so small?

The sea will rescue us from this predicament.

One reason is that the seas are cultivated in a very primitive fashion; indeed, they are hardly cultivated at all. More than 95% of our food is raised by animal husbandry or organised cultivation of plants; the rest is obtained by hunting or by gathering what grows wild. But virtually all that we take from the seas is obtained by hunting, with only a tiny contribution from fish farming and other deliberate cultivation techniques.

In agricultural terms, fishing is a very primitive industry, but there are more basic reasons than this for the shortage of sea food. The sea bed is a splendid storehouse, continually fed by decaying material, which is rich in phosphorus and nitrogen. Unfortunately, however, most of this material never reaches the surface where it could do some good. The coastal areas are more fertile, as are some off-shore regions.

In a few parts of the world—usually on west coasts in fairly low latitudes—the prevailing winds and currents combine to move the surface layers of water out to sea and to bring deeper water, rich in nutrients, near to the surface, where photosynthesis and food production occur. This upwelling has produced the immensely rich fishing grounds off the coasts of California, South West Africa and Peru, but the total area involved is less than a thousandth of the oceans.

The sea is not really a larder, it is a desert floating on a compost heap. It is sometimes supposed that the oceans could be made more productive by the use of fertilisers. Small-scale experiments of this kind have been moderately successful in coastal waters, but it is doubtful whether large-scale fertilisation of the oceans would be justified by increased catches.

A more speculative possibility is the sinking of a nuclear reactor as a source of heat to produce convection currents moving the natural fertiliser to the surface. This procedure might well be successful, but it is fraught with difficulties and, in the present climate of opinion about radioactive pollution, it is unlikely to be attempted.

Technical difficulties are not the whole story. The present catch from the world's fisheries could make a substantial impact on protein deficiency, which is the major cause of malnutrition throughout the world, but it is not being used for this purpose.

Fish flour, made by drying and grinding whole fish and removing the fat, is a bland, tasteless material, but would be a valuable food supplement. It is, however, not widely used, partly because of aesthetic objections to the consumption of fish guts—though these objections have never restrained the rich from enjoying their whitebait or oysters.

The world's largest fishery, producing more than 10% of the entire catch, lies off the coasts of Peru and northern Chile. None of the crop is used to relieve malnutrition in Latin America. A little is used as human food in Japan, and the rest goes to Europe and the United States to help in the intensive production of pigs and poultry.

One of the most striking effects of modern technology has been to widen the gap between the rich and poor nations. There is little profit in feeding the hungry—or, as the poet might have said, man's inhumanity to man makes countless thousands starve.

Wandering Continents

In some green island of the sea,
Where now the shadowy coral grows
In pride and pomp and empery
The courts of old Atlantis rose†.

Masefield was one of the many poets inspired by the legend of a submerged continent. Plato's account (the basis of all subsequent versions) is fanciful, but it now seems that the legend has a grain of truth.

The island of Thera, to the north of Crete, has deep deposits of volcanic ash, useful in making waterproof cement. Excavations to obtain this material needed in building the Suez Canal uncovered a city, which further study has revealed to be a Minoan counterpart of Pompeii; Plato attributed the cult of the sacred bull to the Atlanteans. Deposits of ash on the sea bed indicate that a volcanic eruption (about four times more powerful than that of Krakatoa in 1883) occurred in about 1450 BC. Clouds of ash, earthquakes and tidal waves destroyed towns on the Aegean Islands, killed cattle, and brought the Minoan civilisation to an end. There was a land that sank beneath the Atlantic, but not in a day (like Plato's Atlantis), or even in a million years. This was Rockall Bank, of which only a tiny islet now shows above sea level. Samples of rock from the submerged mass have been found to be mostly granite, showing that Rockall Bank is a slab of sunken continental rock, and not a volcanic structure.

Mr Sullivan's massive but readable book‡ explores many fascinating questions about the Earth's structure and history, beginning with one that can be found on any map of the world: why do the outlines of western Africa and eastern South America match so closely?

During the nineteenth century, a few scientists speculated on catastrophes (usually associated with the Flood) that might have torn the Earth's surface apart. The shapes of the continents were shuffled like a jigsaw puzzle, to show how they might once have been joined. The first to attempt a comprehensive theory of continental drift was Alfred Wegener, a German meteorologist and explorer. Until about three hundred million years ago, he asserted, there was only one continent, which he called Pangaea. Fragments had broken away at various times, up to a million years ago, and drifted apart leaving the oceans in the spaces between. The Mid-Atlantic Ridge, a chain of mountains mostly submerged, but visible in Iceland and the Azores, represented debris left behind when Europe and America parted company. The mountains of western America were made by the wrinkling of the Earth's crust as the continental mass moved westward.

---

† From John Masefield 1932 *Fragments* in *Collected Poems* (London: Heinemann), reproduced by permission of The Society of Authors as literary representatives of the Estate of John Masefield.
‡ Walter Sullivan 1977 *Continents in Motion* (London: Macmillan).

Wegener assembled a lot of evidence, including similarities in rocks, fossils and animals found in regions now widely separated, but which had probably once been joined together. Lemurs are found only in East Africa, Madagascar and the countries on the other side of the Indian Ocean. Fossil ferns of a distinctive kind have been found in India, Australia, South America, South Africa and Antarctica, suggesting strongly that these lands were once joined. Wegener's ideas had a hostile reception, mainly because geologists could not imagine continents sailing through the solid crust of the Earth. There was also distrust of ideas which brought together many separate sciences such as geology, meteorology, biology, oceanography, and even physics. Continental drift was therefore dismissed as unscientific.

Wegener died in Greenland in 1930, still an outsider treated with disdain by the scientific establishment. Today his theory is respectable and no one challenges his ideas, though many interesting problems remain to be tackled. The general picture is that the continental masses, about 20 miles thick, are still moving at a rate of a few inches a year. The forces that drive them are not so mysterious as they seemed in Wegener's time. For example, a block of ice can be shattered with a hammer, but glaciers flow gently down mountain sides. A piece of toffee can be snapped by a quick blow, but will yield gently when chewed. The continents sit on the mantle—a layer of rock which extends from below the crust to a depth of about 2000 miles, where the molten core begins. The material of the mantle can flow if the movement is slow enough. Since this material is continually heated (by the radioactivity of uranium and other constituents), it supports convection currents essentially similar to those generated in a coffee percolator. Molten material from the mantle rises through mid-oceanic cracks, spreads sideways under the continents and eventually descends into deep oceanic trenches. So the continents don't need to be driven; they ride passively on the mantle material as it circulates.

Some of these processes can be seen at first hand. In south-western Iceland the Mid-Atlantic Ridge has, through repeated eruptions, risen above sea level—as did the new island of Surtsey a little to the south in 1963. The Red Sea is another site of volcanic activity, probably signalling the beginning of a new ocean. The validity of these ideas could be tested if samples of the mantle were available. It is remarkable that we know more about the rocks of the Moon than about the mantle which makes up most of our own Earth. Of course no drill will go through 20 miles of continental rock, but there are places in the ocean where the crust is only about four miles thick. An ambitious project to drill through this relatively thin crust began at Mohole off the coast of Mexico in 1958, but was abandoned, mainly because of political troubles, in 1966. Samples of the mantle will be gathered eventually, for the technology is already available. Meanwhile, the Earth's structure is becoming more clearly known. Ingenious researches, which Sullivan describes, suggest that the Earth's surface is composed of about six large plates and a few small ones, riding on the hot plastic material of the mantle brought from below. At the boundaries between the plates there is earthquake activity; this is because the neighbouring plates are usually moving at different rates. The movement, resisted by friction, builds up stresses which are relieved at intervals by earthquakes, allowing the two sides to regain their proper positions.

This process happens regularly (though not always on a large scale) along the San Andreas Fault in California. Slipping between the two sides of the fault produced the San Francisco earthquake of 1906, and further disasters are predicted. There is, however, some prospect of controlling earthquakes. Fault systems might be lubricated by pumping in water to reduce friction and to encourage movement in small tremors rather than large earthquakes. Stresses might also be relieved by nuclear explosions in uninhabited places on the fault.

Well! It isnt my fault!

This process happens regularly ...

The boundaries between the plates forming the Earth's crust are potentially useful as sources of heat. Most of the houses in Reykjavik in Iceland use water heated by the lava which is never far from the surface. Geothermal energy is already used to raise steam for electric power generation in several countries and is capable of supplying a useful fraction of the world's energy.

Sullivan's narrative, clearly written and lavishly illustrated, tells how an understanding of the wandering of the continents helps in the search for oil, metals and diamonds. His story illuminates the interplay of science and technology in other ways too. Much of the basic scientific work that he describes was undertaken to find ways of distinguishing (from great distances) between earthquakes and nuclear explosions, or to provide data relevant to submarine warfare. The new science of geophysics is both the product and the source of technology.

David Knight, who teaches the history of science at Durham University, discusses the transformation of Wegener's ideas from heresy to dogma during the 1960s†. The progress of science may be described in various ways—as a logical succession of small improvements, as a response to the needs of society, or as a succession of revolutions separated by periods of consolidation. The structure of scientific revolutions was convincingly and influentially examined by Thomas Kuhn in 1962‡. Decisive changes, he wrote, are brought about by the emergence of new paradigms. Kuhn gave the archaic word an extended meaning as sets of laws, theories, applications and techniques providing models from which spring particular coherent traditions of scientific research. Newton's optics—and his dynamics—embodied paradigms which dominated areas of scientific endeavour for a long time.

Continental drift, says Knight, was another successful paradigm. Its acceptance was a revolution rather than a predictable advance, and occurred because a community of scientists who had been pursuing well established lines of thought and experiment became convinced, rather suddenly, that the new synthesis defined a worthwhile field of research.

† David Knight 1976 *The Nature of Science* (London: André Deutsch).
‡ Thomas Kuhn 1962 *The Structure of Scientific Revolutions* (Chicago: Chicago University Press).

Sums
and
Such

1235813213455891442333 7610

Playing with Numbers

Write down any two numbers. Add them together and write down the answer to make a third. Then add the second number to the third and continue in the same way a few times more. As this series progresses, the ratio of successive numbers will come closer and closer to a particular value which is roughly 1·6 (or, more accurately, 1·61803398 ...). The simplest form of the series, using the numbers

$$1, 1, 2, 3, 5, 8, 13 \ldots$$

was known in about the year 1200 to the great mathematician of medieval times, Leonardo of Pisa, sometimes known as Fibonacci, and is usually named after him. Fibonacci's *Liber Abaci* ('Book of the Abacus') was the first textbook to advocate the use of the Arabic numerals familiar to us today. Until this revolutionary innovation, arithmetic was a grim business—try multiplying XIX by LVII! Fibonacci introduced European scholars to the ideas of the Hindus and Arabs, and laid the foundations of the study of algebra.

Arithmetic was a grim business ...

The Fibonacci number, approximately 1·618, has some curious properties. If we write it as *F*, then it is not difficult to verify that

$$F - 1 = \frac{1}{F}$$

and

$$F + 1 = F^2.$$

F has properties interesting to artists, architects and engineers, by whom it is sometimes known as the Golden Section or Divine Proportion. If a rectangle is drawn with F as the ratio of its sides, and a square is cut off from one end, the rectangle which remains will have its sides in the same ratio and the process can be continued indefinitely. The ratio is said to be particularly pleasing to the eye, and has been exploited by one or two artists, including Salvador Dali. It turns up unexpectedly in relation to the growth of trees and the arrangement of leaves around plant stems. There is even a Fibonacci Association, with its own quarterly journal full of scholarly mathematical reports describing new properties of the series.

Some mathematicians devote themselves to obviously useful activities, employing symbols and abstract ideas to make models of reality and to produce results of practical value in science or engineering. Others regard their subject as a branch of logic, not necessarily connected with tangible creation. A few devote themselves to the apparently trivial, but really quite profound, task of studying the properties of numbers themselves.

How many nine-figure numbers can be made, using each of the digits one to nine? It would of course be possible to write them all down and add them up, but this approach would occupy a nimble pen for about six months. The problem can be solved in a couple of minutes by observing that the first digit can be chosen in nine different ways. When this has been done there are only eight possibilities for the second digit, seven for the third and so on, giving a total of $9 \times 8 \times 7 \times 6 \times 5 \times 4 \times 3 \times 2 \times 1 = 362\,880$ different nine-figure numbers. Ogilvy and Anderson, the authors of an introduction to number theory[†], go on to ask: how many of these numbers are exactly divisible by three? They give the simple answer: all of them. The reason, as they explain, is that any number will divide exactly by three if the sum of its digits is divisible by three. The sum of the digits one to nine comes to 45, and therefore any number containing these nine digits will be an exact multiple of three. The proof of this rule, which Ogilvy and Anderson expound in detail, is not difficult to follow.

Their excursion leads along many interesting byways, pausing occasionally to admire splendours of scenery or man-made eccentricity. Many spectacular short cuts and other devices used in tackling apparently intractable problems are shown to be based on the relatively simple basic properties of numbers. The calculating prodigies of earlier times were seldom able to give a clear account of their methods, but they certainly used aids of this kind. It is significant, Ogilvy and Anderson point out, that no mathematical prodigies have been born during the twentieth century, although it is unlikely that anyone would be very excited by demonstrations of unusual ability in mental arithmetic in an age when electronic computers are daily performing tremendous feats of figuring. A few achievements of this sort are mentioned, and reference is made to oddities such as Skewes's number which, though it has a real significance, is so big that not all the paper in the world would contain it.

[†] C S Ogilvy and J T Anderson 1967 *Excursions in Number Theory* (London: Oxford University Press).

The authors are, however, not only concerned with curiosities. Their object, in which they succeed very well, is to show that the simple sequence of numbers conceals problems which—though providing exercise for the most able mathematical experts—can be explained in terms familiar to spectators with only an elementary understanding of the subject.

Trying to get Numbers to be Random Enough

A professor of physics who recently accepted a contribution of £250 000 from the Science Research Council observed (with the transparent honesty of all his kind) that the experiments which had attracted the grant would be of no practical use.

Whether his judgment was inspired by pride or by humility the mere printed word does not reveal. The scientific establishment, however, would applaud his conclusion, and rightly so because, strictly, all science is useless. Technology, which often inspires or accompanies science (and sometimes grows out of science), is a different story.

Naturally enough there are degrees of uselessness and even a generous patron can be pushed to the edge of incredulity. When hard pressed, scientists will sometimes admit that their work might—in special circumstances and in a small way—be useful to someone.

Dr Jacques Dutka, a mathematician of Columbia University, used a computer to calculate the square root of 2 to a million places—actually a million and eighty-two, perhaps because he was slow on the switch. For practical purposes, no one ever needs to know this number to more than the three or four places that every schoolboy remembers, but Dutka proposes that his million-digit answer will be useful as a source of random numbers, needed in preparing trials of new drugs or fertilisers, and also in many other enquiries where statistical impartiality is supposed to be preferable to human prejudice or emotion.

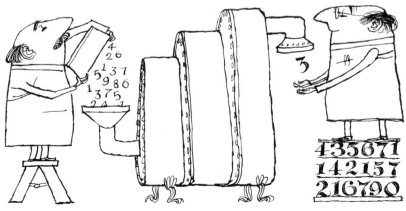

Random numbers can be produced by electronic devices or by simple machines.

Random numbers can be produced by electronic devices (such as ERNIE) or by simple machines. The telephone directory is not of much use because, for obvious reasons, it contains too many ones and twos, but not enough eights and nines. Even the more skilfully made tables sometimes appear to go wrong.

Sir Ronald Fisher, the eminent statistician (who, among his many other activities, advised the tobacco manufacturers on alternative explanations for the prevalence of lung cancer among smokers), once prepared a large table of random numbers, but discovered that the list contained too many sixes, so he crossed out a few to restore the balance. Two other experts discarded 10 000 numbers because they did not like the look of the figures produced when an assistant was operating the machine.

Suppose that we make a random-number machine out of an old gas meter, which would not be very difficult. Recording the numbers that it delivers, we find, on the first page, two runs of four consecutive sixes. Reason advises that the sequence 6666 is just as likely to occur as 3584 or any other combination, but common sense tells us that there is something odd. We then add a gadget to the machine so that any run of four or more identical digits is discarded before the results are printed out.

The engine starts up again and produces a book of four million random digits which, for convenience in reading, are printed in blocks of four. A mathematician buys the book and complains that is is not random at all. He expects to find a run of four identical digits once in every thousand blocks. There are, he explains, ten thousand numbers between 0000 and 9999, among them ten which (like 9999) are composed of identical digits.

Trying to get numbers to be random enough.

He will therefore hope to find about a thousand sequences of this sort among the million four-figure numbers that he has bought, but actually he discovers none. Who is right? The mathematician could resort to the Trade Descriptions Act to get his money back.

Dr Dutka has the ambition to calculate π to a million figures before he retires from the manufacture of large numbers. The result certainly ought to be random, but there are suspicious circumstances. π has already been calculated to ten thousand places; anyone with a half crown to spare could have bought the lot (along with a great many words) as long ago as 1957.

The ten thousand figures do not pass reasonable tests for randomness. It is said that each of the ten digits occurs about a thousand times, but there is a run of six consecutive nines near the beginning, and one block of a thousand figures contains the sequence 7777 three times. The million-figure version will be an interesting curiosity, but not many people will buy it as a list of random numbers, because the experts fear that numbers like π, completely untouched by human hand, are not really random enough.

Blaise Pascal and Probability

If gambling ever becomes respectable enough to have its own patron saint, Blaise Pascal will be the odds-on favourite.

Although known to history as a theologian and philosopher, Blaise Pascal's most interesting work was done in science and mathematics. It included the first analysis of probability, leading to results which have found ever-widening applications in industry, medicine, agriculture and many branches of research but—fortunately for the bookmakers and pools promoters—they are not properly understood by the gambling fraternity.

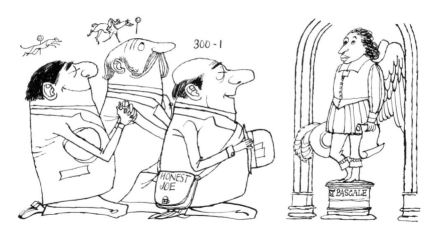

Blaise Pascal will be the odds-on favourite.

Pascal was a child prodigy. His father, not wanting his son to be overworked at school, employed tutors but confined the boy's education to languages. Mathematics, he thought, would be too dangerous. This restriction aroused Pascal's curiosity and he began discovering for himself.

At the age of 12, he cut a triangle out of paper. Folding over its three sides, so that they met on the base line, he showed that the sum of the three angles was equal to 180 degrees. This demonstration would not be accepted by mathematicians as a rigorous proof, but it does show remarkable ingenuity.

The study of Euclid naturally followed, and a lot of new ground was explored in geometry. In 1640, at the age of 17, Pascal made the first calculating machine, the forerunner of today's great arithmetic engines. By this time the ill health that was to persist for the rest of his life was well established, with dyspepsia by day and insomnia at night.

Not surprisingly, he pondered deeply on the misery of man and the prospect of salvation in the hereafter. One night, in an agony of toothache, he sought distraction by recalling some famous problems on the cycloid—a graceful curve formed, for example, by the motion of a point on the rim of a moving wheel. Galileo had discovered some of its properties and, in a flash of intuition, suggested its use as an arch for bridges—a bright idea which, now supported by the laws of mechanics, is widely used to this day.

After a while, Pascal's toothache abated. Seeing an omen in the unexpected relief, he bent his mind to the problem in earnest, and after eight days had the solution sewn up pretty thoroughly.

More often, however, his torments were assuaged by literary efforts, to which we owe the immortal *Pensées* and the *Provincial Letters*. His occasional ventures into science produced brilliant results nevertheless. Noting Torricelli's refutation of the old proverb 'Nature abhors a vacuum,' he had glass tubes, 40 feet long, prepared in Rouen, filled them with wine or water and persuaded his brother-in-law to carry them up and down the Puy de Dôme.

Nature, it seemed, did not abhor a vacuum quite so much at the top of a mountain. Pascal himself made further tests on the less exacting heights of the Tour St Jacques in Paris. This work revealed the nature of atmospheric pressure and led to the enunciation of the laws of fluid pressure in general, which now form the basis of all hydraulic machinery.

Antoine Gombaud, Chevalier de Méré, was a gambler who struck a bad patch in 1658. For quite a while he made a modest living by wagering that he could roll at least one six in four throws of a dice. Seeking to add variety to his repertoire, he decided that the chance of a double six in 24 throws of two dice should be just as good. Experience soon showed that there was something wrong with his reasoning, for he swiftly lost most of his profits.

He complained bitterly to Pascal of the scandalous condition of arithmetic, which was self-contradictory and sometimes not even true. Pascal found (what can now be proved with little trouble by an intelligent schoolboy) that the Chevalier's first wager had a slightly better than even chance of success, and that in the long run he would grow rich; but that the second variation, with two dice, had less than an even chance of gain.

Another problem posed by de Méré was the division of a prize for an unfinished game of chance. Further study of these matters led Pascal to construct a general theory by which questions of chance and probability could be resolved without actually writing out all the possible outcomes and counting them up. Here he made use of a remarkable arrangement of numbers known today as Pascal's triangle; it was familiar long before to Omar Khayyam who did not, however, appreciate all of its powers.

Each number in the triangle is the sum of those to the left and right immediately above it. Together, they have some useful properties. Bridge

```
            1
         1     1
       1     2     1
     1     3     3     1
   1     4     6     4     1
 1     5    10    10     5     1
```

players should memorise the first few lines of the triangle to help in guessing the distribution of the cards. Suppose for example, that four cards of a given suit are outstanding, with no clues as to their positions in the two opposing hands. The fifth line in Pascal's triangle gives, in condensed form, the probabilities of each of the conceivable patterns. These are, from left to right: 0–4 (probability $\frac{1}{16}$); 1–3 ($\frac{4}{16}$); 2–2 ($\frac{6}{16}$); 3–1 ($\frac{4}{16}$); and 4–0 ($\frac{1}{16}$).

Bridge players should memorise the first few lines ...

In other words, there are six ways of finding an even distribution 2–2, but ten ways in which the cards may lie unequally between the two hands. The exact probabilities, which may be found in books on the mathematical theory of bridge, are a little different, but the triangle gives results good enough for practical purposes.

The theory of probability is firmly based on the mathematical equivalent of common sense, but many people think that they know better. Gamblers, in particular, have an innocent belief in the hoary old fallacy known as the law of averages. If black turns up several times in succession, they say in the casinos, put your money on red the next time. In fact, the chance of red is still exactly one half, however many times black may have appeared (assuming, of course, that the wheel is in good order, as it always is).

When black came up 26 times in succession at Monte Carlo (on 18 August 1913), none of the clients made much gain, but the casino did very well. After the fifteenth black, everyone started to bet on red, many of them doubling their stakes after each spin because they thought that yet another black was impossible.

The casino proprietors would, of course, do quite well even if all their clients were mathematicians. The rules of the game are arranged so that the house takes 1·35% of the stakes over a sufficiently long period of time (this is the figure at Monte Carlo; elsewhere the house takes generally rather more).

In the short term, an individual gambler may take money away from the tables, particularly if he has enough capital to ride out a sequence of failures. But the house has bigger reserves than any client and, in the long run, will claim a steady 1·35% which amounts, in Monte Carlo, to an annual return of 125% on the invested capital represented by the casino.

Statisticians, actuaries and scientists of every kind rely heavily on Pascal's work, but the people who might profit most from it are the hardest to convince. That is why there are lots of hard-up gamblers, but no hungry bookmakers.

The Cross that Every Schoolboy Knows

The man who put the cross of St Andrew into the arithmetic books died more than 300 years ago. Whenever a schoolboy writes $2 \times 2 = 4$, he uses the multiplication symbol first introduced by William Oughtred, Rector of Albury in Surrey, who was one of Europe's foremost mathematicians. (The printer, if he be not learned in the byways of mathematical history, will sometimes use the letter x, but the authentic symbol is a cross.)

Oughtred was born and educated at Eton College, where his father was the pantler (defined in Dr Johnson's dictionary as 'the officer in a great family who keeps the bread'). After three years as an undergraduate, and eight years as a Fellow of King's College, he left Cambridge in 1603 for the vicarage of Shalford in Surrey, where his ministry lasted until he moved to Albury in 1610. He was (according to John Aubrey)

'He slept but little ...'

... a little man, had black haire and black eies (with a great deal of spirit). His head was always working. He would drawe lines and diagrams on the dust. His eldest son ... told me that his father did use to lye a bed till eleven or twelve a clock, with his doublet on, ever since he can remember. Studyed late at night, went not to bed till 11 a clock, had his tinder box by him, and on the top of his bedstaffe, he had his ink-horn fix't. He slept but little. Sometimes he

went not to bed in two or three nights, and would not come down to
meales till he had found out the quaesitum.

Similar feats of abstraction and concentration are not unknown even
among contemporary mathematicians, but are unusual in an amateur, such
as Oughtred was. He seems to have devoted more effort to his hobby than
to his vocation.

His neighbouring ministers said 'that he was a pitiful preacher; the
reason being that he never studied it, but bent all his thoughts to the
mathematiques. But when he was in danger of being sequestered for a
royalist, he fell to the study of divinity, and preacht (they sayd) admirably
well, even in his old age.'

Oughtred's first mathematical works, written when he was in his
early twenties (but not published until many years later), were on the
construction of sundials, which were important instruments in the days
before watches or portable clocks. ('And then he drew a dial from his poke;'
As You Like It II, vii.) His first major work was the *Clavis Mathematicae*, or
'Key of Mathematics', which appeared in 1631 after considerable pressure
from his friends and pupils (among whom was Christopher Wren). This was
a work of immense importance, and was one of the books which influenced
Newton as an undergraduate some 30 years later.

Its merits were twofold. Firstly, it gave a short, clear summary of
arithmetic and algebra which were, in those days, difficult subjects.
Secondly, Oughtred recognised the need for standard symbols to
identify mathematical quantities and operations. He invented 150 of these
symbols, and some of them remained in common use long after his death,
though only the multiplication sign is universally known today. This
innovation appeared first in 1618 in an anonymous appendix (now
attributed to Oughtred) to the first English translation of John Napier's
book on logarithms.

Napier of Merchiston regarded his logarithms as an aid to
trigonometry, and therefore to navigation. Oughtred showed how useful
they could be in other ways. He was quick to see the labour-saving
possibilities of logarithms in multiplication and division and, in 1622, he
invented the slide rule. In this early form of computer, two logarithmic
scales moved side by side, providing a quick and simple means of adding two
logarithms, and therefore of multiplying the corresponding numbers. The
first model used circular scales, and a later version with straight sides looked
more like the slide rule so common in recent times.

Oughtred was a great teacher, but he did not favour the use of slide
rules and other instruments by students. When one of them asked why the
invention of the slide rule had been concealed for so many years, Oughtred
answered:

That the true way of Art is not by Instruments, but by Demonstration: and that it is a preposterous course of vulgar teachers, to begin

with intruments, and not with the Sciences, and so instead of Artists, to make their Scholers only doers of tricks, and as it were Juglers.

This argument is still going on, even though engineers and scientists have abandoned the slide rule in favour of the pocket calculator, which does even cleverer tricks.

Oughtred died at the age of 86 in 1660. Legend has it that he heard of King Charles II's return to the throne, called for a glass of sack to drink a loyal toast and expired from intense emotion. As the Restoration occurred on 29 May, and Oughtred died on 13 June, the celebration must have been formidable indeed.

Puzzles and Paradoxes

Scottish mathematicians have never been noted for their high spirits. John Napier of Merchiston, the inventor of logarithms, dabbled in alchemy but otherwise comported himself with the utmost solemnity. John Aubrey tells how the Professor of Geometry in Cambridge 'travelled into Scotland to commune with the honourable lord Napier of Marcheston about making the logarithmicall tables.'

Another historian, who was present at their meeting in Edinburgh, 'brings Mr Briggs up into my lord's chamber, where almost one quarter of an hour was spent, each beholding the other with admiration, before one word was spoke.'

William Jack, a Glasgow graduate, was more articulate, serving from 1870 to 1875 as editor of *The Glasgow Herald*, and afterwards for 30 years as Professor of Mathematics. More recently, Professor A C Aitken of Edinburgh has given astonishing demonstrations of memory and mental agility as a lightning calculator.

On the whole, the study of mathematics is a serious business in Scotland, but Mr T H O'Beirne of Glasgow University has spent many years showing that sums can be fun, even to those with no more than schoolroom mathematics†.

... attributed to Bede in one of his less venerable moments.

Some of the most interesting problems are derived from ancient or trivial sources. The anguish of the three zealous husbands and their frivolous

† T H O'Beirne 1965 *Puzzles and Paradoxes* (London: Oxford University Press).

wives, trying to cross a river with a boat that holds only two people, appeared in a French collection which is still in print, though it was first published in 1612. This teaser is attributed also to Bede (in one of his less venerable moments) and used by Alcuin, another English monk, to encourage his pupil, the Emperor Charlemagne. Mr O'Beirne explores all the variations with missionaries and cannibals, wolves and dogs, goats and cabbages. The answers become steadily more complicated, eventually appearing in large diagrams much resembling the Blackpool illuminations. The logical solutions of playing a mouth-organ to distract the cannibals or bringing a polythene cabbage to divert the goat are, of course, not admitted.

During the Second World War a great deal of scientific effort was diverted into the study of the 12-coin puzzle which seeks to find, by only three weighings on a balance, which of the dozen is bad, and whether it is over- or under-weight. Scorning the traditional methods, involving nothing more elaborate than the teeth—or a hard table-top and a good ear—the boffins on both sides grappled with this exercise to the detriment of their duty. The guns had hardly stopped firing in 1945 before solutions poured into the editorial offices of the mathematical journals. The counterfeit coin problem is now thoroughly licked, with detailed solutions for as many as 120 coins (in five weighings), and general rules for even more spectacular feats of assay.

The first men faced with the problem . . . probably settled for an approximate solution.

The first men faced with the problem of dividing a gallon of ale, using only a three-pint and a five-pint jug, probably settled for an approximate solution. Mathematical exactitude involved triangle-mesh nets, drawn on graph paper ruled with sloping lines. Even more decorative (suitable for wallpaper, suggests Mr O'Beirne) is the sample of 25-point geometry—a network of lines, letters and triangles which sparkles like a masterpiece of op-art when closely viewed. These abstractions turn out to have useful properties in the design of race tracks—a possibility which is, it must be regretted, little appreciated by the stewards of the Jockey Club.

The true mathematician is never idle. One of Mr O'Beirne's researches started at a Christmas party when an exploding cracker threw out the challenge: 'A farmer sells 100 head of cattle for £100. A cow sells for £5, a sheep for £1, a pig for 1s. How many of each were there?' The answer is not very hard to find (19 cows, one sheep and 80 pigs), but the puzzle has a surprisingly long history. Abu Kamil, a ninth-century Arab, gave his readers 100 drachmae to purchase 100 fowl of three types: ducks at five drachmae, hens at one drachma, and sparrows at 20 for a drachma. Three hundred years earlier, Chinese scribes were busy with cocks, hens and chickens, while Indian teachers preferred geese, cranes and peacocks.

Abu Kamil quickly gathered an audience when he said: 'I am acquainted with a type of problem which proves to be engrossing, novel and attractive alike to high and low, to the learned and the ignorant.' The same promise might be held out to readers of Mr O'Beirne's book, but the author must know that a change is coming.

The computer which was turned loose on Euclid a little while ago astonished its keepers by producing an entirely original proof for the *Pons Asinorum* (a baffling proposition about the angles at the base of an isosceles triangle) in about six lines. This was better than Euclid could do, but the solution might have been found by any schoolboy during the last century, for it involved only the simplest ideas. When the machines get into their stride they will produce books of tantalising puzzles that no one will be able to solve—no one, that is, except another computer.

General Knowledge Arithmetic

'Here's an interesting case,' said the Dean of Admissions at a respectable American university not long ago. His colleagues, who had already spent many hours sorting chaff from chaff, listened while he read school reports and referees' assessments: English very good, mathematics adequate, French terrible, Latin poor, fond of travel, aiming at a military career, interested in politics. After pondering for a while, they came to an unfavourable conclusion and were passing on to the next candidate when they learned that they had rejected the 18-year-old Winston Churchill.

The Civil Service Commissioners (who allowed Churchill into Sandhurst at the third attempt) were kinder, although not quite so generous as the headmaster of Harrow who, a few years earlier

> ... took a broad-minded view of my Latin prose .... I wrote my name at the top of the page. I wrote down the number of the question '1'. After much reflection I put a bracket round it thus '(1)'
> ... It was from these slender indications of scholarship that Mr Welldon drew the conclusion that I was worthy to pass into Harrow†.

Not every duffer turns into a Churchill, but no one wants to be too hard on the pupil of merely average attainments, and much effort is devoted to helping him over the hurdles. It seems, however, that examinations are still too stiff. The Educational Institute of Scotland were recently moved to complain that science papers for the Scottish Certificate of Education increase in difficulty every year. The taxpayer, who keeps armies of scientists in comfort and supplies them with accelerators, computers, and electron microscopes for the sole purpose of making science ever more complicated, might complain if matters were otherwise, but the schoolmaster faces facts and wonders how to help his pupils through their examinations.

In the old type of science, a teacher observed some time ago, pupils could gain marks by knowledge of bookwork, but the new courses demanded some power of reasoning, which militated against the ordinary candidates. These complaints came at a bad time, for the teaching of science is now in the revolutionary phase which occurs about once in each century. The ice calorimeter and the tangent galvanometer survived in the schoolroom long after they were consigned to the dustbin of science, but now they are gone even from the textbooks.

Some time in the twenty-first century the more radical educationists will start complaining about the persistence of synchrotrons, semiconductors and such like old-fashioned rubbish in the school curriculum, but there is

† From Winston Churchill 1947 *My Early Life* (London: Odhams Press).

not much hope of any immediate relief for the hard-pressed science teachers.

The mathematicians have done rather better, to judge from a recent report on an experimental examination designed for the Certificate of Secondary Education. This particular parchment is intended for the pupil of average ability, who has been happily browsing on bookwork all these years. The new test is certainly different, for it ignores Euclid, the binomial theorem, the differential pulley, and the other familiar constituents of examination questions. Stranger still, the paper contains no questions, but only 21 'situations'. One of these, it must be admitted, concerns a man walking from P to Q at $4\frac{1}{2}$ mph and back at $2\frac{1}{2}$ mph, but the rest make no concession to nostalgia, either in the problems which are to be tackled (after studying the situations), or even in the form of the permitted answers, which are to be written in little boxes. There is no scope here for a Churchill, and grave discouragement for the lad who chooses to express himself in pages of calculation before offering the wrong answer.

The truth is that the mathematicians of England have neatly solved the difficulty of catering for the average pupil, by making the test so strange and formidable that the weaker brethren will not go near it at all. This is a Draconian remedy, which might have been avoided by a little subtlety.

William Butler, teacher of writing, accounts and geography in ladies' schools and in private families, offered a gentler solution in his *Arithmetic Questions* which appeared at the end of the eighteenth century. Here the pill of reasoning or calculation was liberally coated with jam in the form of general knowledge and historical commentary. Question 40, for example, is concerned with addition and subtraction, but begins:

> Union of England and Scotland. Caledonia was the ancient name of Scotland, whofe king, James VI, fucceeded to the throne of England, on the demife of Queen Elizabeth, which aceffion produced the union of the two crowns in 1603.

The teacher goes on to discuss the union of the Parliaments in 1707:

> This union was not, however, accomplifhed without much trouble and fome art and intrigue; and upwards of £20 000 were, it is said, diftributed among the leading men in Scotland.

After some further background information, comes the crunch:

> How long have Scotland and England been united this prefent year 1799? Anf. 92 years.

The diligent student soon masters addition and subtraction, passing on to sums about millet ('An efculent grain, chiefly ufed among us in puddings ... a noble sudorific ... what are 58 pounds and $\frac{1}{4}$ of millet worth at $6\frac{3}{4}$d per pound?'), and pausing to reflect (question 86) that Virtue alone is true nobility. Maximilian, Dr Knott and Thomson the poet are quoted in support of this proposition, summarised in Pope's well known lines, that

Worth makes the man, and want of it, the fellow:

The reft is all but leather or prunello. Prunello is a kind of ftuff of which the gowns of clergymen are made. What is the value of $65\frac{1}{2}$ yards of prunello at 5s $3\frac{1}{4}$d per yard?

Although he was writing mainly for ladies, Butler was no pussyfoot. His definition of milk ('A well known fluid') is supported by a wealth of anatomical detail and paediatric advice. He shows real enthusiasm for strong drink, making no distinction between 'whifky' and 'whifkey,' but advising knowledgeably about other tipples:

Wine of Lebanon. Hofea speaks of this wine as very fragrant. chap. xiv, ver 7 ... In two hogfheads of this wine, how many gallons, quarts, and pints?

He shows real enthusiasm for strong drink ...

If today's mathematicians succeed in frightening away all the examination candidates, they should sit down and prepare a new edition of Butler, which could perhaps put them in business again.

The Living World

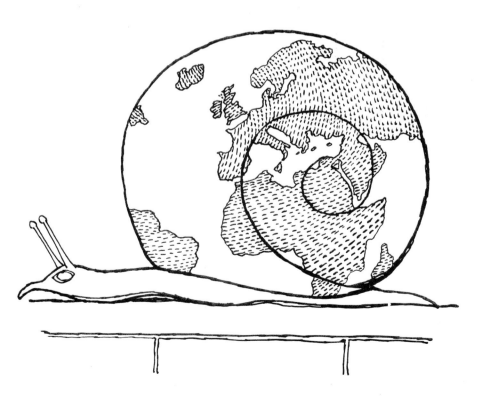

Stranger on the Shore

A plague of plastic bags washed up on various beaches recently provoked complaints which were addressed to the naval authorities, in the belief that the unwanted material had been discharged from the nuclear submarines, whose presence in British waters has been the subject of much discussion.

The culprits were not in uniform. The submariner is trained from an early age to go about his business unobtrusively. The last thing that he would do is to discharge rubbish in a plastic bag likely to float for miles and to make itself conspicuous on the shore. Closer inspection of the bags would have probably revealed a connection with farming, building construction, or one of the other industries where the advantages of cheap, light and durable containers are appreciated.

... provoked complaints which were addressed to the naval authorities.

Many kinds of rubbish can be discarded in the confident expectation that they will disappear if merely left alone for a long enough time. Tin cans will rust, and paper and cloth will rot, helped by ingenious bacteria able to live on cellulose.

An acre of woodland may produce as much as a ton of fallen leaves in a year, but the ground level does not noticeably rise. The leaves remain intact only until the first wet day, after which they are attacked by bacteria, fungi, and micro-organisms of other kinds. The process of decomposition is hastened through the activities of slugs, snails, earthworms and insects. Before long the dead leaves have been converted into humus which, incorporated into the surface soil, helps to nourish future generations of trees and other plants. Twigs or even fallen trunks are dealt with in the same efficient way. Beetles and other insects invade the wood, making tunnels for smaller creatures, by which the cellulose and other constituents of the wood are eaten or reduced to dust.

Polythene is a British invention, developed by ICI and first produced on a small scale in 1938. Like many other plastics, it is made by the joining together (or polymerisation, to use the technical term) of a large number of small and relatively simple molecules.

When the basic materials are rearranged to form very large molecules, usually in the shape of long chains, the characteristic properties are produced. Polythene is made from ethylene, a substance with the chemical formula C_2H_4 found in coal gas and natural gas, and which is quite easily made in large amounts in other ways.

In the original ICI process, the ethylene molecules were made to join up by bringing the gas to a temperature of about 200 °C and a pressure of about 2000 atmospheres. Other processes have since been developed giving polythene of even better properties by polymerisation at more normal temperatures and pressures.

Polythene is a remarkably good electrical insulator, and has mechanical properties which make it useful in a great variety of domestic and industrial applications. In molten form it has a low viscosity and can therefore be made into thin film. This film is waterproof and can be sealed merely by heating.

For these reasons, polythene is an excellent material for the plastic bags now used as containers for confectionery, fertilisers, and many other commodities. Unfortunately, polythene bags are all too durable. They have no value as scrap, and cannot be burned without producing unpleasant smoke and fumes. When emptied and discarded they carry the danger of suffocation, both for babies and for cattle. They are not appetising to any of the creatures which help to keep the countryside tidy by feeding on garbage.

Life on Earth is a cyclical process. Living creatures come and go, returning their raw materials to the ground. Water, oxygen, carbon dioxide and other basic chemical substances appear over and over again, sometimes in the living body, sometimes in the atmosphere, the seas or the Earth's crust. This satisfactory and economical arrangement is now being disturbed by the development of materials which do not come into equilibrium with the biological environment.

Materials such as minerals, cellulose, and living tissues of plants or animals, have been used for a long time—and without drastic modification—to support the requirements of civilised life. When chemists and engineers apply their skills to the production of materials quite unlike anything found in nature, they may be creating a gigantic waste disposal problem for the future.

Remember the aboriginal chief whose subjects subscribed to give him a new boomerang on his sixtieth birthday, and who spent the rest of his life trying to get rid of his old one.

A Sense that is not to be Sneezed at

'Hellish dark,' said James Pigg, 'and smells of cheese.' The huntsman had been sitting with Mr Jorrocks discussing prospects for the next day's sport. His report on the state of the weather was somewhat inaccurate because, confused by the darkness and fumes of brandy, he mistook a cupboard for the window, demonstrating that smell is sometimes a keener sense than sight.

The eye has a remarkable performance as an optical instrument, but its powers can be surpassed or greatly extended by devices such as the microscope and the spectrometer. Light signals can be detected, recorded and analysed in very precise ways, but the application of corresponding techniques to the study of smells has hardly begun. Lacking understanding of the basic processes involved, it is not possible to make an instrument with a performance even approaching that of the nose.

In the course of its evolution, the human nose has undergone changes which have not improved its appearance or functional efficiency. In the horse, the dog, and many other animals, the air follows a fairly straight path into and out of the lungs. In man the change to an upright posture and the great growth of the brain have been accompanied by anatomical adjustments, among which the hairpin bend in each nostril is a conspicuous and inconvenient feature. The dog can sneeze without much difficulty, but in man the out-going blast of air is frustrated by the narrow twisted outlet and has to escape through the mouth.

The human nose, for all its ugly shape, is a well designed mechanism to filter, warm and moisten the air on its way to the lungs. The olfactory bulb, which is the basic organ of smell, is not separated from the outside world by anything corresponding to the lens of the eye or the eardrum. The direct connection of the olfactory nerve fibres to the brain is a reminder of the great importance of smelling power at earlier stages in man's evolution.

A keen sense of smell is no longer essential for survival. Consequently, the sensitivity of the human nose is often surpassed by that of insects, dogs and other animals.

Experiments have shown that the threshold of concentration at which various fragrant substances can be detected is always lower for a dog than for a man. Further study has shown that, for some chemicals, a dog is about a thousand times more sensitive than a man, but for many other substances the dog's threshold of perception is more than a million times lower. A dog's nose is far more sensitive to substances such as narcotic drugs than is almost any scientific instrument.

Where the dog has a thousandfold advantage, he is probably smelling the same smell and using rather similar equipment to do it. The dog's nose is designed to give free access to the air, and his brain is probably better able

The sensitivity of the human nose is often surpassed by . . . dogs.

(from experience, if from no other circumstance) to deal with the signals that reach it. When the sensitivity is a million times better, the dog is probably using powers that are not found at all in the human subject. We might explain this further by suggesting that the smell mechanism covers a wider range of odours in the dog or allows certain substances to be detected in a way that man cannot emulate. However, these speculations avoid the main problem, which is to explain how the process of smelling really works, either in a dog or in a man.

Recent experimental work uses sensitive analytical instruments to identify and measure the components of a smell. A practical application involves the correction and study of vapours from the human body. For this test the subject is put into a closed glass tank through which a rapid stream of purified air is passed in order to collect the breath and other emanations, which are concentrated and analysed. Thirty or more components are usually found, but the pattern is by no means constant, even for a single individual.

The 'manpack personnel detector' now in service with the United States Army uses an electronic nose to detect the presence of an unseen enemy. Carried on a soldier's back, the detector takes an air sample through a flexible pipe and sounds an alarm when it identifies 'sub-microscopic agents or particles given off by humans.'

'To fmell . . . to ftrike the noftrils.' It would be hard to improve upon Dr Johnson's summary. Even after 200 years we do not have a clear idea as to what goes on between the sniffing of an odorous material and the sensation of smell conveyed to the brain.

In recent times there have been four serious attempts to provide a scientific explanation for the mysteries of olfaction. All of the theories which have been offered use the same starting point: the interaction between molecules of the material in question, and the cells in the small region of the nose where the sense of smell resides.

... uses an electronic nose to detect the presence of an unseen enemy.

The first explanation, based on the plausible suggestion that chemical reaction is the key, was then explored in a long series of experiments carried out more than 40 years ago by G M Dyson, a chemist interested in perfumes. He began by studying phenyl mustard oil, a synthetic compound chemically related to a material extracted from mustard seeds. He then made a whole series of chemical compounds, all slightly different, by adding chlorine, bromine, iodine, and other atoms at various places in the original molecule. The smell of each of the new substances was recorded in the hope that some simple relation might be found with the chemical structure. The chemical properties of the various compounds were also studied in the hope of finding a correlation.

Dyson was not deceived by some partial successes in this quest and observed: 'No chemical data, either from the viewpoint of reactivity or chemical structure, will give us the key to the rational, quantitative interpretation of odour phenomena.'

Secondly, a curious theory, proposed 30 years ago (and soon discarded), suggested that the olfactory nerve cells in the nose gave off infrared radiation. If the appropriate chemical substance happened to be passing at the time, this radiation would be strongly absorbed, thereby cooling the nerve endings and giving rise to the sensation of smell. The basic objection to this proposal is that a transfer of heat between the nerve cells and the surrounding air can only take place if there is a difference in temperature. In practice, however, it is found that the sense of smell is still intact even when the air is at the same temperature as the nose. It is difficult also to see how very tiny changes in temperature, corresponding to the extremely low concentrations at which many pungent smells can be detected, could be distinguished from the random changes in temperature produced even by the passage of pure air through the nose.

A third theory, first proposed in 1961, suggests that the sensation of each of the main kinds of smell (such as camphor, ether, peppermint, and a few others) is associated with vacant sites of characteristic shape on the surface of the olfactory organ. Most of these sites are envisaged as circular or elliptical depressions, but some are likely to be more complicated. The appropriate smell will be evoked when a molecule of the right shape drops into one of the vacant spaces. Some support has been given to this theory by the finding that, for a number of synthetically prepared chemicals, smell is correlated with the shape of the molecule. There is, however, no explanation so far of the process by which the correctly shaped molecule generates a nervous impulse when it has dropped into place, nor of the mechanism by which it is ejected after the smell has been recorded.

The fourth theory, which is also quite plausible, stems from the observation that the region in the nose associated with the sense of smell has a distinctive brown or yellow colour, due to the presence of an olfactory pigment. This pigment is found in relatively large amounts in animals which have a highly developed sense of smell. Colour in any object is associated with the presence of electrons which can absorb energy rather easily from an incident beam of light, thereby passing to states of higher than normal energy. The coloured patch inside the nose is not likely to be excited by light, but a similar effect might be produced by the absorption of energy from chemical reactions or from neighbouring molecules.

A few years ago, R H Wright of Vancouver suggested that an electron in a molecule of olfactory pigment might be encouraged to jump (and, in doing so, to generate a signal which could be passed along the olfactory nerve to the brain) by the proximity of another molecule which had the appropriate natural frequency of vibration. The effect invoked here is resonance, as demonstrated in the legendary tale of the singer who could shatter a wine glass.

Wright's theory is not universally accepted, though he has made a number of calculations and experiments which support it in quite a convincing way. In studying the properties of the olfactory pigment, he concluded that it might have a chemical structure similar to that of vitamin A. It has been known for some time that rats deprived of vitamin A appear to lose their sense of smell. Two Australian scientists who examined the olfactory organs in dogs and cows found clear signs of the presence of vitamin A or of some similar substance. They then gave large doses of the vitamin to a group of 56 people who had no sense of smell, even though they were free from disease or injury. In 50 of the patients the power of smell was restored partly or completely.

At one stage during the Second World War the success of new radar equipment in British aircraft was camouflaged by the rumour that pilots improved their night vision by munching carrots. Vitamin A occurs in the visual pigment of the eye and can be made in the body with the help of a substance found in carrots. Keener vision is quite attractive, but most people are probably happy with the sense of smell that nature provides, and it will certainly be a long time before bunches of carrots are given away with bottles of perfume.

Counting the Calories

... I confefs it will afford me a fingular pleafure if I can prove, by experiment, that a pleafant and varied diet is equally conducive to health, with a more ftrict and fimple one.

William Stark was a pioneer in the scientific study of nutrition. Born in Birmingham in 1740, he graduated as Master of Arts at Glasgow University in 1758, and studied medicine in Edinburgh and London before taking the MD of Leiden in 1767. He lived at a time when knowledge of diet and digestion was dominated by superstition and ancient authority.

Foodstuffs were thought to be made out of the basic elements of air, earth, fire and water, linked with the properties of heat, cold, dryness and moisture. A person's temperament could be sanguine, phlegmatic, choleric or melancholic, and the food that he ate had to be adjusted accordingly.

Foodstuffs were thought to be made out of the basic elements ...

But times were changing. The spirit of scientific enquiry was growing, especially in chemistry and physiology. Stark was one of the first to appreciate that a satisfactory diet might be constructed on a rational basis. He planned a long series of experiments to test the response of the body to various foods, and (he hoped) to show that a varied and palatable intake was just as healthful as any of the bizarre diets advocated by cranks or even by experts. He was impressed by advice of an eminent politician:

> Dr B Franklin of Philadelphia, informed me, that he himfelf, when
> a journeyman Printer lived a fortnight on bread and water, at the
> rate of 10 lbs of bread per week and that he found himfelf ftout and
> hearty with this diet.

Stark tried living on a diet of bread and water, with a dash of sugar,
for ten weeks. By the end of this experiment he was feeling pretty ill and
was suffering from scurvy. He changed to a more liberal diet, with milk and
meat, and recovered his health. After a month during which he lived
entirely on puddings (made from flour, fat and water), he began to study the
effects of lean and fat meat and continued with a diet of honey, bread and
Cheshire cheese until he died in February 1770.

Stark was a brave man, with ideas ahead of his time. A century
later, he would have made his experiments on animals and lived to a ripe
age himself. In 1770, the scientific understanding of nutrition was just about
to be achieved. By 1790, the experiments of Lavoisier and Laplace had
shown that the process of respiration and digestion were basically similar to
combustion: air was taken in to support the oxidation, and water and carbon
dioxide were discharged as waste products.

The Frenchmen showed that the amount of heat produced by a
guinea pig (living in a calorimeter) was proportional to the amount of carbon
dioxide which the animal breathed out; in other words, to the amount of
carbon that had been converted to carbon dioxide in the body. Later
experiments of a more exact kind, using human subjects, showed that the
amount of heat produced by the body as a result of all the processes of
digestion, respiration and exercise, was equal to the heat of combustion of
the food eaten by the subject; that is, the amount of energy obtained by
burning the food as efficiently as possible in a calorimeter.

The body burns its food in a very sophisticated way, but the subtle
forms of energy used in biochemical processes or in muscular and mental
exertion are eventually degraded to heat. It is for this reason that we can
express food values in heat units (calories), though the calorie used for this
purpose is a thousand times bigger than the calorie used in other branches of
science.

An average man uses about 500 calories during eight hours of sleep;
1200 calories during eight hours of work (assumed to be not entirely
sedentary but not physically exhausting); and about 1500 calories in other
activities, including walking, recreation, and merely sitting around.

The fuel intake of this imaginary individual should therefore be
about 3200 calories per day. The diet must, of course, fulfil certain other
requirements as regards the relative amounts of protein, carbohydrates, fats,
minerals and vitamins. More than half of the world's population has to exist
on a diet which is deficient in quality or quantity (or both), but in wealthy
countries the problem is rather different.

Most people in Britain have an adequate diet because, as William

Stark suspected (but did not live long enough to prove), a satisfactory balance can be obtained in many different ways. Well fed people do not count the calories—and do not need to—unless they are overweight. Although scientific experiments seem to account for the whole of the heat of combustion of the food which is eaten, they cannot be absolutely exact. A discrepancy of 1% would be quite acceptable in the laboratory, but the average person takes in half a ton of food in a year, and 1% of that is 11 pounds.

The simplest way to lose weight is to eat less. It is easy enough to lose several pounds in a month or two without serious discomfort. Unfortunately, most people don't begin to take dieting seriously until they have 20 or 30 pounds to lose. When this stage has been reached, there is no substitute for expert advice from the physician and, perhaps, the dietician. Persons of normal girth can continue to observe Stephen Leacock's advice:

> If you like nitrogen, go and get a druggist to give you a canful of it at the soda counter, and let you sip it with a straw. Only don't think that you can mix all these things up with your food. There isn't any nitrogen or phosphorus or albumen in ordinary things to eat. In any decent household all that sort of stuff is washed out in the kitchen sink before the food is put on the table†.

† From Stephen Leacock 1910 *Literary Lapses* (London and New York: The Bodley Head and Dodd, Mead and Company).

Artificial and Natural Foodstuffs

The popular imagination (than which nothing is more influential or more irrational) recognises a clear distinction between natural food and artificial food.

Natural food is the genuine stuff, grown on soil uncontaminated by insecticides and other new-fangled poisons. If fertilisers are used they must be organic materials known to our forefathers; anything that comes in a sack is a nasty chemical, repugnant to lovers of pure food. Chickens, whether intended for the pot or for the production of eggs, must be allowed to run about freely, enjoying a natural environment rich in dirt, germs, parasites, and other benefits.

... a clear distinction between natural and artificial food.

Artificial food is adulterated by the presence of chemical fertilisers made in factories, sprays and powders applied to control natural pests, and various other materials that nature never intended. Artificially raised chickens are allowed nothing but clean food and water in adequate amounts, and are kept in warm surroundings all their days.

The enthusiasts are convinced that natural food has a higher nutritive value, and all the scientific evidence in the libraries and laboratories will not change their minds. The attitude that nature knows best extends to the other materials taken into the body. Proteins, carbohydrates, and such like things are not really chemicals, provided that they have been raised by highly principled farmers. Vitamins are good for everyone; glucose is a super-charged variety of sugar; and aspirin is an essential aid in dealing with the headache, the common cold, and any unclassified ache or pain. Most other chemicals are popularly classified as drugs, a word connoting, at the best, dangerous medicines but—more often—noxious substances which are agents of addiction and degradation.

To the scientist the distinction is not so clearly drawn. Many useful drugs are natural products found in plant or animal tissues. On the other hand, many of the foodstuffs that we eat contain chemicals which in other circumstances would be rightly classified as poisons.

When the first banana appeared in a grocer's shop in London in 1632 the pharmacists raised a loud complaint, protesting that the new fruit was a drug, probably poisonous, and should be sold only under the supervision of professionally qualified experts. More recently, the Hippies of San Francisco put banana skins in their pipes and smoked them, but once again found no convincing evidence of toxic constituents.

... put banana skins in their pipes and smoked them ...

But nutmeg (and mace, which comes from the same plant) contains substantial amounts of myristicin, a powerful hallucinogenic drug. 'Take a quantity of maces and chew them well,' wrote Richard Banckes in 1525, 'hold them there awhile and that shall loose the fumosity of humours that rise up to the brain, and purge the superfluity of it.' In more precise terms, half a gram of nutmeg (not much more than a pinch) will often produce a detectable response. A teaspoonful, as snuffed by South American Indian witch-doctors, produces violent hallucinations which are sometimes fatal. Myristicin occurs also in celery, parsley, and a few other vegetables, though not at dangerous levels.

In 1964, Dr Ivan Stewart of the Citrus Experimental Station at the University of Florida, made the interesting discovery that synephrin is present in lemon juice. Synephrin and a number of related compounds are used in nasal sprays to relieve the symptoms of the common cold; the use of hot lemonade in the same connection may therefore be soundly based. Synephrin belongs to a class of compounds known as sympatheticomimetic amines; biochemists give long names to almost everything, reserving the catchy titles for notorious compounds such as DNA and LSD. Sympatheticomimetic amines are quite abundant in cheeses of the more pungent kinds, such as Camembert and Stilton. These amines can produce unpleasant effects, but are normally put out of action promptly by an enzyme known as monoamine oxidase in the body.

In 1963 there were many reports of a distressing illness in patients undergoing treatment with certain tranquillisers. Some astute physicians noticed that the attacks were always associated with the eating of cheese. After that the picture became clearer. The tranquillisers in question were all materials which interfered with the action of the monoamine oxidase enzyme; the relation of this property to their tranquillising action was not fully understood. When the patients took cheese and tranquillisers within a period of an hour or two, the toxic substances were unrestrained and the unpleasant effects quickly followed. With this explanation the problem was solved and very few cases have occurred since.

Enzymes are useful substances, and the body could not function without them. Occasionally they display unusual properties, as do those in lima beans and peach and apricot kernels, where enzyme action converts relatively harmless compounds into hydrogen cyanide, otherwise known as prussic acid. The amounts produced are fortunately below the toxic level, although often well above the legally prescribed limits for prussic acid in foodstuffs.

It is hard to support the widespread opinion that ordinary, everyday victuals are indisputably good and that man-made chemicals in food are bad; food is a mixture of chemicals, not all of which are necessarily good for us.

The Sound of Medicine

Discoveries and inventions are seldom made at the right time. The first clinical thermometer was made in about the year 1600, but the instrument did not achieve regular use in medicine until well into the nineteenth century. The modern type of thermometer, with a short stem and a constriction allowing the patient's temperature reading to be retained even after the bulb has been taken out of his mouth, is little more than a century old.

Radioactive isotopes became available in useful amounts 30 years ago, but would have been eagerly accepted in science, industry and medicine at any time during the past half century. X-rays provide an unusual example of a discovery which could hardly have been made any earlier (because the necessary technology had not been developed) but was, on its appearance, immediately pressed into service by people who knew exactly how to use it.

The stethoscope could have been invented at any time during the last 2000 years, for it is based on very simple technology adapted to a long-standing clinical need. Hippocrates, the Father of Medicine, knew that sounds emerging from inside the body could be of diagnostic value. One technique which he recommended was to hold the patient by the shoulders and shake him, and to listen for splashing sounds. A gentler method of investigation was to apply an ear to the chest and listen for the heartbeats.

Dr Rene-Theophile-Hyacinthe Laennec was a physician, more than 150 years ago, at the Necker Hospital in Paris. This famous institution is still there; the French, like the British, have no problems about old hospitals—they just keep on using them. Describing his experiences with 'pectoral audition,' Laennec wrote:

> I was consulted in 1816 by a young lady who presented the general symptoms of a heart disease and with whom the application of the hand and percussion gave poor results owing to stoutness. The age and sex of the patient forbidding the type of examination of which I have just spoken, I remembered a well-known phenomenon of acoustics: if the ear is applied to one end of a beam, a pin-prick is most distinctly heard at the other end.

There is a legend that Laennec was reminded of this phenomenon of acoustics by some small boys whom he saw playing with a plank of wood in the courtyard of the Louvre. Whatever the source of the inspiration, he wasted no time in exploiting it. Rolling up a few sheets of paper to make a tube, he applied one end to the patient's chest and the other end to his ear. He found that the heartbeats were heard more clearly and distinctly than ever before. Before long the paper tube was replaced by a hollow cylinder of beech wood with suitably shaped ends, to which the inventor gave the name of 'stethoscope.' By other physicians it was known as the Medical Trumpet or Pectoriloquy. The modern type of stethoscope, supplying sound to both ears, developed gradually during the nineteenth century, and came into common use some 60 years ago.

'The age and sex of the patient forbidding the type of examination of which I have just spoken ...'

Laennec and his pupils established the value of the stethoscope in the diagnosis and study of many afflications of the heart and lungs, making a significant contribution to the physician's repertoire. In recent years the technique has been extended in a variety of ways. The cathode-ray tube (which provides a vehicle for television pictures, but also has more serious uses) allows heart sounds to be recorded permanently on photographic film or paper. Noises produced in other parts of the body can also be studied in a systematic way.

Photocraniography is the study of pulsating sounds heard when a stethoscope is applied to the skull. These sounds come from the blood vessels in the head and can sometimes be used to investigate disorders of the circulation. A common method is to use an electronic stethoscope (which turns the sound into an electrical signal) coupled to a cathode-ray tube or a tape recorder.

Abdominal rumbling is more often a cause of embarrassment than a source of enlightenment, but a serious study of bowel sounds was made as long ago as 1905 by W B Cannon, a celebrated American physiologist. He used an ordinary stethoscope, but sophisticated electronic techniques have since been developed and have yielded a good deal of fresh information. The process of digestion generates a wide spectrum of sounds; most of them, fortunately for the bystanders, are inaudible to the unaided ear, but they are music to the gastroenterologist.

The sensitive methods now available for the detection and recording of feeble sounds offer considerable scope for the ingenious listener. A physician in Yugoslavia recently recorded pulsations and murmurs from the thyroid gland in the neck. No one seems to have made any acoustical study of the kidney and its associated plumbing or of creaking joints. There is no

... music to the gastroenterologist.

shortage of experimental material, for the human frame (like Prospero's island) is full of noises, sounds, and sweet airs.

Electrodes in the Brain

Dr Jose Delgado stood in the bullring. His adversary pawed the ground, snorted, put his head down, and charged.

Dr Delgado pressed a button on the small black box which he carried in his hand. The bull faltered and slithered to a stop. When he charged again, a few seconds later, he was halted in the same unusual way. The box in Dr Delgado's hand was a radio transmitter, sending signals to a small electronic implant in the bull's brain.

... faltered and slithered to a stop.

Dr Delgado became a professor at the Yale University School of Medicine and is no longer seen in the bullring, although he made exciting and important experiments involving the application of modern electronic techniques to basic problems in physiology and psychology.

The idea that various functions of the body are associated with particular regions of the brain is an old one, based on correlations between clinical observations and post-mortem findings. Systematic mapping of the brain began in Berlin in 1870, with the work of Fritsch and Hitzig, whose technique was to apply weak electrical impulses to various parts of the exposed brain of a living dog. They confirmed the intuitions or predictions of earlier investigators, who believed that muscles on the right side of the body were controlled from the left side of the brain, and vice versa. Their work showed also that a wide range of bodily activities could be provoked by electrical stimulation of the cortex, which forms the roof of the brain.

In 1874, Dr Robert Bartholow, Professor of Medicine in Cincinnati, was the first to study the effects of electrical stimulation on the human brain.

In the following year, Richard Caton of Liverpool was able to detect small electrical signals produced spontaneously in the brain cortex of rabbits and monkeys. By that time it was known that one particular region, known as the visual cortex, was associated with the sense of sight; Caton found that exposure of the eye to bright light greatly increased the electrical output from the visual cortex. In 1929, Hans Berger of Jena succeeded in picking up electrical signals from the human brain cortex, using metal electrodes placed on the scalp. Improved versions of his electroencephalograph are now used routinely in hospitals.

All of this work was concerned with the outer layers of the brain which are, of course, more accessible to the needle or electrical detecting equipment.

The brain stands up quite well to the insults of the experimenter. In particular, it does not feel any pain, once the overlying bone and skin have been removed. Needle electrodes, which can be inserted into almost any part of the brain without doing appreciable damage, were used by many investigators. The disadvantage of this technique is that it allows exploration of the brain only while the animal is under anaesthesia with part of its skull removed.

Some 40 years ago, W R Hess of Zürich made an important advance by implanting electrodes which could be left in position permanently, with wire connections passing through the skull. His original work was done with cats, allowing experimental studies to be made on the conscious animal going about its normal activities. Work of this kind is now being carried out with human subjects at many laboratories, most of them in the United States.

A few years ago I spoke to a quiet middle-aged man—a reprieved murderer—who was helping the neurologists in a famous American medical

The brain stands up well to the insults of the experimenter.

school. When he took off his hat, two 50-pin sockets, of the kind commonly seen in electronic instruments, were visible on the top of his head. Inside his skull were 100 platinum electrodes which had been introduced under anaesthesia, and precisely positioned at points of interest in the brain. With this remarkable collaborator—who enjoyed perfect health and travelled freely around the country—a team of neurologists and psychiatrists were mapping out the regions in the brain associated with all of the major emotions. More recently, work of the same kind has been extended to allow the study of memory, the regulation of blood pressure, and many other functions.

These electrical studies of the brain involve more than idle curiosity. Some investigators claim success in the diagnosis—and even in the treatment—of mental illness. Others, remembering Dr Delgado's triumph in the bullring, hope to learn more about the biological basis of antisocial behaviour. At this point the prospect becomes somewhat alarming. If behaviour can be controlled by electrical means, the electronic psychiatrists of the future may have powers which even the boldest reformers would hesitate to use. Those who are most knowledgeable about current research suggest that the brain has built-in defensive mechanisms, providing resistance to manipulation of the personality. Enough is known about the techniques and results of brainwashing to suggest that this optimism is unfounded.

When Habit is Addiction

Jean Nicot was the French Ambassador to Portugal just over 400 years ago. His despatches contained enthusiastic accounts of a herb, widely used by Indian tribesmen, which he had found on his travels. In his honour the plant was named *Nicotiana*.

The first variety introduced into Europe was used mainly in pipes (although Catherine de Medici, Queen of France, soon became fond of snuff), and another species, *Nicotiana tabacum*, became the main source of the world's tobacco, a position which the hardy plant holds to this day.

Catherine de Medici soon became fond of snuff.

Tobacco was first esteemed for its medicinal qualities. Nicot related some remarkable cures:

> ... one of the Cookes of the sayde Embassadour hauing almost cutte off his thumbe, with a great chopping knyfe, the Steward of the house of the sayde Gentleman ran to the sayde *Nicotiane*, and dressed him therewith fiue or six tymes, and so in the ende thereof he was healed.

A detailed account of the new drug was given by Nicholas Monardes, a physician of Seville, in 1571, and translated into English (under the title 'Joyfull newes oute of the newe founde world') by John Frampton in 1577.

Tobacco, like alcohol and other potent substances, soon passed out of the hands of the physicians. King James I was not very pleased. His *Counterblaste to tobacco* expressed violent objection, and he did not stop at mere words. To his High Treasurer he wrote on 17 October 1604, complaining that tobacco was

> ... excessivelie taken by a number of ryotous and disordered Persons of mean and base Condition.

He therefore instructed that the import duty should be increased from 2d to 6s 8d per pound. The King's objections to tobacco were mainly emotional, but it was a long time before the hazards of smoking were to be defined in a more rational way, and the enjoyable effects of smoking have only recently received any detailed scientific explanation.

The relationship of smoking to lung cancer and other diseases affecting the heart and lungs is now well established, although there are naturally a few people who challenge the majority view. What is more remarkable is that millions of people continue to smoke, regardless of expert opinion and official advice. Clearly they must find some benefit to compensate for the risk of fatal disease.

The enjoyable effects of tobacco are associated with the nicotine which is an important constituent of the smoke. Some of its properties are well known and easily verified. In many people the smoking of a single cigarette will increase the pulse rate by ten beats per minute, and will reduce the temperature of the skin of the hand by several degrees.

The action on the mind is more complicated. Nicotine enters the brain rather easily when tobacco is smoked, chewed, or snuffed. Once there it releases noradrenaline, a substance manufactured in nerve cells, and which is rather similar to the adrenaline made elsewhere in the body to be delivered to the bloodstream in times of excitement or emergency. The release of noradrenaline in moderate amounts from the brain relieves tension, promotes concentration, and gives a general feeling of well being. Amphetamine—sometimes known as Benzedrine—produces rather similar effects. It seems that tobacco is a drug of addiction and that threats or persuasion will not have much effect on its devotees.

The mechanism linking smoking with lung cancer is not known with certainty, but is generally believed to involve benzpyrene, a substance which is found in tobacco smoke and is certainly capable of producing skin cancer in mice. Attempts to produce a safer cigarette are sometimes based on the action of filters in removing benzpyrene and other dangerous constituents. It is by no means clear that this method is completely effective, but there are other possibilities.

The temperature in a burning cigarette can reach 840 °C, a level at which relatively harmless constituents of the smoke may be converted into benzpyrene. The lesser hazard of pipe and cigar smoking is often attributed to the lower temperature at which the tobacco is burnt. A home-made cigarette burns at a slightly lower temperature than the manufactured article, but the temperature of combustion falls to below 650 °C if the tobacco is loosely packed.

Chemical tests, and experiments on mice, suggest that the low-temperature cigarette is considerably safer than the standard commercial product—news which will perhaps stimulate the sale of those fascinating little machines used by do-it-yourself enthusiasts in times of war or rationing.

Grass is Good for you

Why do we need meat? Mainly because we are not plants, and because (in the nutritional sense) we are singularly inefficient animals.

A student whom I once met across the table summarised the situation very neatly. Faced by one of those threadbare questions that we use in the ritual exercise of teasing the young, he had to compare the eye and the camera. 'No comparison at all,' he said. 'The camera is made of tin but the eye is made of meat.'

Unfortunately, all animals (including humans) are utterly dependent on plants for the protein needed to build and maintain their bodies. The merest weed performs feats of chemical construction that no chemist can imitate. A green plant takes in sunlight, water, carbon dioxide and simple chemicals (found in the soil or in seeds), and builds them into carbohydrates (roughly speaking, starch, sugar or oil) and into proteins.

The plant is not very efficient. Half of the Sun's energy is useless for making food and much of the remainder is reflected from the leaves instead of being absorbed. Science can do nothing about this waste, which is just as well, because if plants did not reflect light, they would all look black.

All in all, a farmer's crops use 1% of the energy that falls on them from the Sun in making carbohydrates and proteins. King Nebuchadnezzar (who was, incidentally, a good farmer) lived briefly on leaf protein, but we usually let the plant go on growing until the protein has made its way into the seeds or tubers.

Flour and potatoes are often condemned as starchy and therefore (in popular belief) fattening, but they also contain useful amounts of protein. People who live largely on wheat products often take in three times as much protein as a well fed European, but still suffer badly from malnutrition because some proteins are more nourishing than others.

The commonest proteins, such as wool, hair and horn, are useless as foods. A mixture of leaf proteins could be made into serviceable victuals, but the experiment is seldom tried. In their natural state, leaves are indigestible by man because of the amount of cellulose that they contain.

The human stomach, which is a rather simple chemical plant, cannot cope with cellulose. Cattle, on the other hand, are equipped with more and bigger stomachs (sometimes reaching 50 gallons in capacity) inhabited by bacteria capable of turning the cellulose into carbohydrate, producing enough methane to light a street lamp in the process.

Potentially nourishing plants are commonly fed to animals, partly because our delicate digestions cannot break down the cellulose, but also for a more important reason. What we really need are not proteins but amino acids, the simpler substances from which proteins are built.

There are about 20 varieties of amino acids, and the human machine needs every one of them. The tissues of the body can make about a dozen (by rearranging other amino acids or perhaps by putting smaller molecules together), but the remaining eight have to be provided at first hand by plants or at second hand by animals which have eaten the plants.

... producing enough methane to light a street lamp.

No single source of protein—not even the over-priced beefsteak—contains all the amino acids essential to man. For this reason, we have to obtain our daily protein from a variety of sources: some from vegetables, some from meat, and some from eggs and milk. In view of the world's protein shortage (which is what food shortage really means, because there is an abundance of carbohydrates), it is unfortunate that animals are such inefficient converters.

Of the digestible protein eaten by cattle, only about 10% is turned into edible food protein; the performance of the maligned broiler chicken is two or three times better than this.

In nutrition, as in so many other aspects of life, the major contribution of technology has been to widen the gap between the rich and the poor. Oil seeds grow well in hungry tropical countries and many of them (such as groundnuts) contain a lot of protein. But after the oil has been squeezed out the residue is discarded or, worse still, used to feed pigs and hens in overfed countries such as Britain.

We have, of course, been hypnotised into the belief that raw red beef is great stuff. It certainly improves the taste of horse radish, but complaints about the high price deserve limited sympathy. We can do without for a while; no one ever died of starvation for the want of a steak.

A Gory Story

Dr Croone told me that at the meeting at Gresham College tonight ... there was a pretty experiment of the blood of one dog let out ... into the body of another ... this did give occasion to many pretty wishes, as of the blood of a Quaker to be let into an archbishop, and such like; but, as Dr Croone says, may, if it takes, be of mighty use to man's health, for the amending of bad blood by borrowing from a better body.

So Pepys wrote in 1666, when blood was still believed to be the vital essence—the distillation of the life itself. Transfusion was recommended, and was sometimes practised, for the relief of insanity, afflictions of the gut, and even marital discord. The results were often catastrophic, for the early experimenters did not know that the body distinguishes cautiously between self and non-self. Consequently, any foreign tissue, whether blood, invading germs, or a heart or kidney, is likely to be destroyed.

Fortunately the rejection process can sometimes be restrained by ingenuity. With blood transfusions, the main problem is the recognition of blood groups, a technique which is basically rather simple (depending on observation of what happens when drops of different people's blood are mixed together) and might, according to Dr Earle Hackett†, have been discovered 300 years earlier had Pepys's scientific friends been more persistent.

... the recognition of blood groups.

Blood-letting was less complicated than transfusion and, on the whole, less hazardous. Opening a vein was a simple way of expelling

† Earle Hackett 1973 Blood: *The Paramount Humour* (London: Jonathan Cape).

demons. When it was decided that diseases were not really caused by evil spirits, blood-letting was prescribed for the relief of congestion, or just because the physicians of olden times did not know what else to do, and did not have today's drugs or electronic aids to choose from. A healthy person hardly notices the loss of a pint of blood (and blood-letting is still a proper treatment in some diseases), but George Washington, King Charles II of England, and many other eminent patients suffered greatly before expiring with their veins almost empty.

Opening a vein was a simple way of expelling demons.

Even the earliest physicians must have noticed that there is not much blood in the body, but a long time passed before the circulation was discovered. Galen, who practised in Italy and Turkey 1800 years ago, thought that blood was freshly made (from food) in the liver, and that it then disappeared among the hungry tissues, just as irrigation water soaked into the dry soil of the Mediterranean countries.

Animal dissection was quite popular and might have revealed the true story, but the Greeks were better at theories than at experiments. Even Harvey, who finally discovered the truth, seems never to have measured seriously a patient's pulse rate. 'In the course of half an hour,' he wrote in 1628, 'the heart will have made more than one thousand beats, in some as many as two, three or even four thousand.'

Dr Hackett (an Irish pathologist working in Australia) deals expertly and entertainingly with folklore and superstition, as well as with science and medicine. The nobles of Castile, he explains, started one myth by claiming to have inherited blue blood. Actually, we all have bluish blood in our veins (easily seen on the back of the hand), but the hidalgos had thinner and clearer skin than the peasants.

Dragon's blood is a resin exuded by an oriental tree; steak does nothing for a black eye; and a horse's pulse rate is only 20 or 30 beats per minute. Robespierre—described by Carlyle as 'the seagreen Incorruptible'—may have been suffering from iron deficiency anaemia (but the cream-faced loon in Macbeth was merely frightened); the engagement ring is worn on the third finger of the left hand because of a mythical vein leading straight to the heart ... Dr Hackett's book is full of interesting stories and asides.

The strangest tale is the true history of blood. Life began in the sea—an ideal place, with food all around and no need for heating or cooling. However, only a very tiny creature can nourish itself by the diffusion of oxygen and other foods through its outer surface, so, as life became more complex, some organisms (such as sponges) developed numerous channels through which the sea could flow, bringing nutrients and removing waste.

Other organisms developed an internal circulation by which oxygen and other necessities, abstracted from the surrounding sea, were distributed among the body cells and tissues. Creatures of this advanced design were more mobile because they carried the necessary environment with them—internally rather than externally.

Some of them left the sea and found how to live on land or in the air. The blood is a relic of the primitive ocean. Blood serum (the clear liquid in which the cells are suspended) has a chemical composition somewhat similar to that of sea water. The content of salts is slightly less than 1%, whereas sea water has rather more than 1%; but the sea is steadily growing saltier and the composition of blood serum is nearer to that of the Palaeozoic ocean of 300 million years ago when life came ashore.

How like an angel ... or, to be more exact, how like a fish.

Man and Molecule

A hen is only an egg's way of making another egg. Richard Dawkins, an Oxford biologist, brings Butler's aphorism up to date by proclaiming† that the animal body is no more than a survival machine for the genes—the units of heredity which probably began in the primaeval soup and are still going strong.

Butler did not understand the mechanism of heredity. Nor indeed did Darwin, whose theory of evolution by natural selection was kept upright by faith and intuition until its rational foundations were completed less than half a century ago. Why do children resemble their parents? How did modern man evolve from the ape-like ancestors which Darwin conferred on him? These were difficult questions for nineteenth-century biologists.

Darwin observed that every generation of animals or plants displays considerable variations in size, colour, and many other less obvious characteristics. He reflected also that plant and animal populations did not increase without limit, as their natural fertility would certainly allow. Putting these commonplace observations together, he concluded that organisms evolve by natural selection. Heritable characteristics which confer an advantage in the struggle for existence will tend to survive, because individuals carrying them will do better in the mating game.

The biologists who accepted Darwin's ideas on evolution through natural selection did not notice an outstanding weakness in the theory. Darwin regarded inheritance as a blending of parental characteristics—a superficially reasonable view supported, for example, by the results of mating between people of different colours. It was left to Fleeming Jenkin, an engineer, to point out that an advantageous characteristic would be diluted by half at each mating and would, after a few generations, be lost. Consequently, evolution by natural selection, depending on the transmission of desirable characteristics through successive generations, would not succeed. The resolution of this difficulty was not finally achieved until 1930, but an important clue was available at the height of the Darwinian controversy. Unfortunately, no one recognised the importance of a report in an obscure journal, over the name of an obscure monk. Mendel's work, which was not appreciated until 1900, showed that heredity works by the transmission of discrete characteristics and not, as Darwin had supposed, by continuous blending. The apparent compromise between parental characteristics occurs only because a large number of factors are inherited and the effect of each of them is small.

From a distance, a half-tone picture cannot be distinguished from a continuous-tone photograph; but the discrete pattern of dots can be seen under a magnifying glass—and so it is with heredity. The picture needs a powerful microscope (and a few other aids), but the mechanism is now fairly clear.

† Richard Dawkins 1976 *The Selfish Gene* (London: Oxford University Press).

The heritable characteristics are preserved by the chromosomes, threadlike structures present in every cell of the body. Man has 46 chromosomes. Each heritable characteristic is associated with a particular segment of a chromosome; these segments (or rather the chemical structures that they embody) are the genes. The genes, says Dawkins, are the lords of creation. Animal (and therefore human) behaviour is regulated by the necessity for survival—not the survival of the species, the community or the individual, but the survival of the genes for which we are no more than temporary lodgings. The genes are made from DNA (deoxyribonucleic acid), a substance which is the key to life, for every DNA molecule is a copying machine, producing faithful copies of itself. Each DNA molecule is a chain of smaller molecules, used like building bricks. These bricks come in four kinds, and can be strung together in an enormous variety of ways. Once put together, a DNA molecule (and therefore a gene or a chromosome) has the capacity to reproduce itself; for example, in growth or in the replacement of our worn-out cells. Each chromosome is really an encoded message, containing not only the instructions for replication, but also a portion of the complete specification of the structure and function of the body to which it belongs. A plan and a copying machine do not make a house; but the DNA molecules also direct the manufacture of proteins from materials lying around in the cells. Proteins make up most of the structural materials of the body (such as skin, muscle and blood cells), as well as the enzymes which regulate the chemical processing of digested food to provide the energy that keeps us alive.

Evolution is not merely reproduction. If all the genes copied themselves faithfully there would be no renewal of the pool of variability needed for natural selection. It has been known since the beginning of the twentieth century that the hereditary material often changes spontaneously. These changes, called mutations, took on a sinister significance with the later

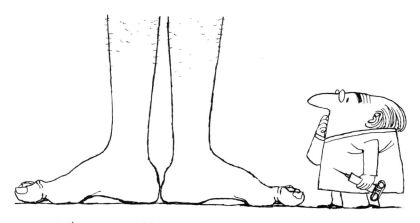

... mutations ... could be produced by radiation or certain chemicals.

discovery that they could be produced by radiation or by certain chemicals; this was the first recognition of man's power to alter his genetic constitution, probably for the worse.

Spontaneous mutations, says Dawkins, are copying errors. The DNA duplicator is not perfect, but the gene machine can cope with the mistakes, exploiting variations which lead to improvements, and phasing out those which are harmful. This, at any rate, is what happens in natural populations. In civilised human communities, such imperfections are not so easily suppressed.

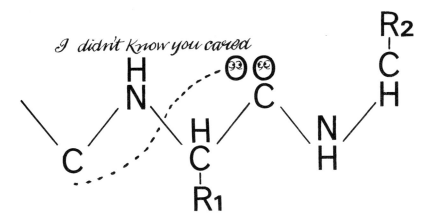

The DNA duplicator is not perfect ...

Having assembled the groundwork, which is straightforward and uncontroversial, Dawkins examines the implications of his model of the animal body as 'a survival machine built by a short-lived confederation of long-lived genes.' Life is short, but genes, reshuffled in each succeeding generation, are almost immortal. This thought leads Dawkins to consider the question: What use is sex? Certainly the survival of the genes does not require the complicated methods of reproduction that animals have developed. Female greenflies can produce fatherless offspring, each containing copies of all its mother's genes. Some simple organisms reproduce by splitting into two, and some plants propagate by putting out runners. In man (as in other animals and many plants), the process is more elaborate. Most of the cells in the human body contain 46 chromosomes—23 derived from each parent. When one of these cells divides, the chromosomes are duplicated. The sex cells, however, divide by a different method to produce only 23 chromosomes, each of which is a patchwork assembly of genes, some from maternal and some from paternal chromosomes. In the process of fertilisation the chromosomes of the sperm cell join with those of the egg to make a new set of 46, programmed with all the characteristics of a new individual.

There are no obvious reasons why this tortuous scheme is the best. It does, however, allow for considerable variation in the characteristics handed on from one generation to the next, since each sperm and each egg has a unique genetic constitution. The important point, says Dawkins, is that a gene can survive through many generations. A chromosome lasts for only one generation before being reshuffled; a gene, being smaller than the fragment involved in the reshuffling, can last a lot longer. Not all genes survive. They are engaged in a game of musical chairs, in which the winners occupy the limited number of vacant niches on the chromosomes of future generations. So the gene is the basic unit of selfishness.

How do the genes express their selfishness? Dawkins considers many aspects of animal behaviour, reviews the currently fashionable explanations, and produces alternative theories based on the selfishness of the genes. He begins with family planning. The human race is threatened with over-population because death rates have been reduced without a corresponding limitation of birth rates. In other species the population is stabilised by starvation, disease or predators—but these controls are not always fully extended, because most animals do not breed to the limits of which they are capable. Sometimes possession of a territory, won in combat, is a licence to breed, and so the weaker males find no mates. Some biologists think that flocks of starlings or swarms of midges are taking a census of their populations, not by counting, but by some sensory mechanism linked to the reproductive system to control breeding: an example of automation in the biological realm.

Other students of behaviour believe that a hen-bird regulates her clutch size so as to maximise the number of surviving offspring in an environment with a limited food supply. Perhaps, says Dawkins (rather unconvincingly) a bird can predict, by judging the abundance of holly berries, or the abundance of her neighbours, whether food is likely to be adequate in the coming spring. If the signs are unfavourable she will lay fewer eggs. Gamesmanship becomes more apparent in the suggestion that, if individuals are reducing clutch sizes on the basis of population estimates, it will be advantageous to make as much noise as possible in the winter roost. Birds will then judge the population to be larger than it is, and will reduce their clutches below the optimum size. Dawkins concludes, plausibly enough, that genes which make an individual have too many offspring tend not to persist because animals containing such genes have less chance of surviving to maturity.

Much of Dawkins's book is concerned with birds and other wild animals, but he offers interesting ideas about human behaviour. The battle of the sexes arises, he suggests, from the fact that in every species sperms are smaller and more numerous than eggs. Consequently, males tend to be promiscuous, since their expenditure of genetic material in mating is relatively small. Eggs are more valuable resources and females therefore tend to be choosy. The female's investment in mating is also greater because she has to provide all the nourishment needed by the embryo until it is mature enough for independent life. For these reasons (and others which Dawkins elaborates), it is to be expected that males will advertise their attractiveness while females can afford to be less conspicuous. This pattern of behaviour is observed in most species, but in modern Western society women are preoccupied with sexual attractiveness while men dress and decorate themselves more conservatively. Dawkins observes that man's way of life is now determined largely by cultural influences rather than by his genes.

Developing this idea further, Dawkins suggests that human behaviour is greatly influenced by units of cultural transmission, which he calls memes. These are ideas, catch-phrases, fashions, techniques, and even tunes. Memes, like genes, spread through a population, though from mind to mind rather than from body to body. The idea of God is a meme, as is the idea of hellfire. These two memes reinforce each other (as genes often do), and assist each other's survival in the same pool. The copying mechanism for memes is not always as accurate as it is for genes. A tune will survive unchanged (if only because it can be written down), but ideas are reprocessed in each brain that they enter; otherwise there would be little work for historians or literary critics. Dawkins does not have a clear answer to this question, but suggests that memes can nevertheless last even longer than genes. Socrates may have no genes alive today, but his meme complex is still going strong. This, like much of what Dawkins says, is a provocative claim. A particular gene (or a consortium of genes) always produces the same result: blue eyes, crinkly hair, or a bushy tail. But do Socratic memes have the same impact on us as they did on Adam Smith or on the medieval schoolmen? The analogy between biological and cultural evolution is fascinating; Dawkins might also have mentioned André Siegfried's essay[†] on a cognate theme.

Dawkins concludes that memes offer more for man's future than genes. Even though we are basically selfish, our capacity to simulate the future in our imagination can allow us to control our cultural evolution: 'We are built as gene machines and cultured as meme machines, but we have the power to turn against our creators.'

† André Siegfried 1965 *Germs and Ideas* (London: Oliver and Boyd).

The End of the Dinosaurs

The Earth, says Professor Ditfurth† (a German psychiatrist), is not an island. Space research and less spectacular studies in laboratories and observatories have revealed much information about our relation to the rest of the universe. Indeed, he asserts with characteristic panache, astronomy has advanced more in the last ten years than in the preceding four centuries. This bold claim is developed with enthusiasm to make an interesting story.

The Moon, he continues, is more than a lump of cold stone. It is largely responsible for the development of life on Earth. The key to this unlikely equation is magnetism. The lodestone of ancient times and the mariner's compass needle are driven by the Earth's magnetic field. Roughly speaking, the Earth behaves as if it had a bar magnet along its axis of rotation, with poles (that is, centres of magnetic influence) at each end. The magnetic north pole is in the Arctic, some hundreds of miles from the geographical north pole, and moves a few miles every year.

As well as guiding navigators, the Earth's magnetic field deflects some of the atomic particles which arrive from the Sun as cosmic radiation and would, if not checked, produce unpleasantly high radiation levels.

Nearly 70 years ago it was found that the Earth's magnetism had been reversed in earlier times. The clue to this unexpected change is trapped in the rocks which, whatever their recent history, retain the magnetic fingerprints impressed on their iron content at the time of their original cooling. If the rocks have not been moved in the interval, their residual magnetism indicates the direction of the Earth's magnetic field at that time.

Some rocks are magnetised in the opposite direction to that of the modern Earth, suggesting that the Earth has reversed its magnetism. The change can be dated, since the age of a rock is usually known from various geological considerations. It appears that the last magnetic reversal was about 700 000 years ago, and that there have been nearly 200 reversals during the last 76 million years.

Why is the Earth's magnetism reversed from time to time? Why indeed is there any magnetic field at all? There is a lot of iron in the core, but the kind of magnetism associated with lumps of iron disappears at high temperatures. Then again, the Earth's magnetism is not constant, since the magnetic poles are always on the move. It therefore seems likely that the magnetism is generated continually by a process subject to fluctuation. If the liquid core and the solid mantle of the Earth are rotating at slightly different speeds, electric currents flowing in the core could produce a magnetic field.

Where these currents and rotations come from is still difficult to explain. Ditfurth observes that Venus, with a size and density not much different from the Earth, has no magnetic field. The reason is, he suggests, that Venus has no moon. The difference in the speeds of rotation of the

---

†Hoimar Von Ditfurth 1975 *Children of the Universe* (London: Allen and Unwin).

Earth's core and mantle can be explained, in a rather complicated way, to be a result of the friction and the consequent slowing down produced by the tides.

These speculations go some way to explain the origin of the Earth's magnetic field, but not why it changes direction now and again. Suppose, however, that a large meteorite hits the Earth; we know that this has happened many times in the past. The impact might jar the liquid core and even bring its rotation to a halt for a thousand years or so, after which it might start up in the same direction or in the opposite direction. During the period when the Earth's magnetic field was interrupted, cosmic rays would strike the Earth in greater abundance.

This change might produce noticeable changes in the living world. All forms of life can be altered by radiation; if the effects are not immediately evident they may be lurking as mutations, waiting to show themselves in future generations. In man most mutations are harmful, but plants and lower animals may benefit through the opening of new lines of evolution. When this happens, advantageous mutations are favoured by natural selection and harmful changes tend to be eliminated. Pursuing this line of thought, Ditfurth turns to a different branch of science.

The ocean floor is made up of successive layers of sediments (sometimes compacted into hard rock) containing the remains of living organisms which once swam in the water. The sediments also contain a great many tiny particles of dust, composed of iron and other metals, which were released into the Earth's atmosphere through the break-up of meteorites. Each of these dust particles is a tiny compass needle which settled on the sea bed in the direction of the Earth's magnetic field.

Using deep-sea drilling techniques developed for oil prospecting, it is possible to obtain a vertical core representing deposits laid down over millions of years. Examination of successive layers under the microscope shows that some species of tiny organisms disappeared (and apparently became extinct) at times when the meteoritic dust particles indicate a reversal of the Earth's magnetic field. The reason, says Ditfurth, is that during the period when the Earth's magnetic field was extinguished, living matter was subject to greatly increased amounts of radiation, leading in some cases to the disappearance of entire species.

This speculation is open to question. Very little cosmic radiation reaches the ocean bed, because it is heavily absorbed in the water. Even on the Earth's surface the cosmic radiation level is kept in check by the atmosphere, which is equivalent to about 30 feet of water. Ditfurth believes, however, that the course of evolution has been greatly influenced by the increase in cosmic radiation (and consequently in mutations) associated with magnetic reversals. He explains that a species which is still developing needs a steady supply of mutations to provide alternative paths for evolutionary improvement, although a species (such as man) which has reached the peak will probably be damaged by further mutations.

The fate of the dinosaurs is interesting in this connection. For 30 million years the dinosaurs—large and small, carnivorous and herbivorous, walking and swimming—dominated the Earth. Then, 200 million years ago, they declined and were supplanted by the first mammals. Perhaps, says Ditfurth, their fate was accelerated by a burst of radiation.

Mark my words! It's a bad sign!

The end of the dinosaurs.

Ditfurth's bold assertions are probably not justified by existing knowledge, but his book is well worth reading, for it gives an interesting review of some important areas of scientific enquiry and illuminates, in language which is not unduly technical, the interplay of experiment and speculation in constructing scientific theories.

People

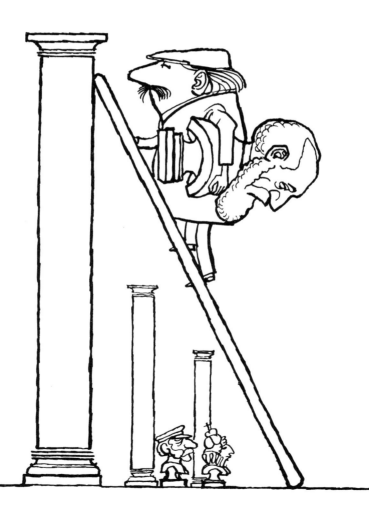

First Signs of Genius

Mr Harvey M Friedman recently became Assistant Professor of Mathematical Logic at Stanford University, California, at the modest age of 19. The report of his appointment added—superfluously, it might be thought—that he was credited with remarkable mathematical ability. Mr Friedman's achievement is unusual, though many mathematicians have shown signs of genius at an early age.

Gaspard Monge was appointed Professor of Physics at a college in Lyons in 1762, when he was 16 years old. Not long afterwards he went to a military academy and applied himself to the theory of fortifications. The object of this branch of the military art was to design earthworks and other barriers so as to avoid direct fire from the enemy.

By applying mathematical ideas to this problem, Monge developed new techniques which were of great practical value, and were indeed treated as military secrets until 1794. The major problem that he solved was a representation of three-dimensional objects by drawings on a flat sheet of paper, using simply a plan (which is a projection of the object onto a horizontal plane) and an elevation (a projection onto a vertical plane). Drawings of this sort are, of course, widely familiar today in engineering and architecture, but Monge was the first to produce them. He went on to greater things in mathematics, took part (as scientific adviser) in Napoleon's Egyptian campaign of 1798, and volunteered to join in the conquest of America when the Emperor's ambitions turned in that direction.

Carl Friedrich Gauss began by correcting errors in the household accounts when he was three years old and (the story goes) discovered at the age of nine how to find the sum of an arithmetical progression. His teacher, seeking time to attend to other duties, told the class to add all the whole numbers from 1 to 100. Almost before the others had started their calculations, Gauss wrote the correct answer on his slate and presented it to the astonished schoolmaster. He had noticed that $1 + 100 = 101$, $2 + 99 = 101$, $3 + 98 = 101$, and so realised that the sum of all the numbers from 1 to 101 was simply $50 \times 101 = 5050$.

Whether this story is true or not, his later achievements were very conspicuous. At the age of 18 he invented the method of least squares, a powerful technique widely used to this day whenever it is necessary to obtain the value most likely to be correct from a series of experimental measurements which are all different.

Gauss was the last man to demonstrate genius in virtually every branch of mathematics. But in his student days he was equally interested in languages and was not sure whether to become a mathematician or a philologist. He made up his mind on 30 March 1796, after achieving a remarkable discovery in elementary geometry.

The Greeks knew how to construct an equilateral triangle or a square, using only ruler and compass. A figure of this kind, with all of its sides equal and all of its angles equal, is known to mathematicians as a regular polygon. It is not difficult to construct regular polygons with five, six, eight or ten sides, using only the simple instruments of ancient geometry, but no one has ever done the job for seven sides. Gauss proved that there is no solution to this problem, but that the task can be done for a polygon of 17 sides.

Gauss's discovery regarding the 17-sided polygon made the first entry in a diary which he kept for 18 years. It contains only 19 pages on which are written 146 brief statements of mathematical discovery, some of which are of enormous importance and others rather difficult to understand. The entry for 19 July 1796 is as follows:

Eureka! num $= \Delta + \Delta + \Delta$.

What he is saying here is that every positive number can be expressed as the sum of three triangular numbers. Three dots on a piece of paper will indicate the corners of a triangle. The same shape can be depicted by six dots (arranged in successive rows containing one, two and three dots) or by ten dots, with a line of four dots forming the base. The numbers 3, 6 and 10 are called triangular numbers, and the series goes on indefinitely, the next members of the series being 15, 21 and 28. For the sake of neatness, the mathematicians add 0 and 1 to the other end of the list. It is not difficult to see that any whole number can be made by adding three triangular numbers; for example, $7 = 3 + 3 + 1$ and $16 = 10 + 6 + 0$.

Of course there have been some clever old men as well. Colonel R E B Crompton fought in the Indian Mutiny and afterwards founded the famous electrical firm which bears his name. He was an engineer of considerable eminence and was proposed many times for the Fellowship of the Royal Society. His sponsors were rebuffed but kept coming back and eventually he was elected in 1933 at the age of 87. The honour which had been so long delayed was not wasted, for the colonel lived on till his ninety-fifth year.

The Poisonous Art of the Borgias

Armchair toxicologists will react with incredulity and indignation to the suggestion (made a few years ago in the *British Medical Journal*) that many English monarchs were not poisioned at all, but died of porphyria, a disease invented by modern physicians.

There is, of course, nothing new about remote diagnosis. The technique is applied more often to historical personages than to common mortals, because the physicians of former days (like the nannies of more recent times) were assiduous in recording anecdotes of their royal charges.

King Charles II had the advice of 14 doctors during his last illness. Although they did not agree among themselves on the nature of the royal malady, their treatments were so vigorous that one of the party (according to Macaulay) 'assured the Queen that his brethren would kill the King among them.'

... they did not agree among themselves ...

The suspicion of poisoning arose as soon as the King's death was announced and 'wild stories without number were repeated and believed by the common people.' Fresh evidence which appeared many years later suggests that he was poisoned, though unintentionally.

In 1961, two American scholars added to the annals of the Royal Society of London (founded by the King in 1660) a note describing his alchemical hobby, which involved the distillation of large quantities of mercury. They concluded that he had died of mercury poisoning, and found abundant support for this conjecture in the contemporary accounts of his illness.

Not long ago a small piece of King Charles' hair was examined by neutron activation analysis and was found to contain about 53 parts of mercury per million—ten times the normal concentration—enough to confirm excessive exposure to the metal, but not to put the diagnosis of poisoning beyond all doubt. This hair was, however, probably gathered several years before the King's last illness.

The poisoners of old were a dedicated race of craftsmen, exploiting the resources of technology to serve their noble masters. The first whose work has been adequately chronicled was Wondreton of Paris, a minstrel before he went into the chemical business in 1384. He was employed by Charles the Bad, King of Navarre, and was commissioned to poison Charles VI of France and several members of the court. He bought arsenic in small amounts from apothecaries and tried to sprinkle it in the royal soup, but was arrested before any harm had been done.

Two hundred years later, Catherine de Medici gave Henry of Navarre a book which had been heavily doctored with arsenic. Her son, Charles IX, read the book first and absorbed some of the poison from the moistened finger used to turn the pages. Suspecting villainy, he consigned the volume to the flames.

The preparation used by Catherine's poisoner was probably based on an Italian recipe. The Christmas greetings which Madonna Caterina Sforza sent to the Pope in 1499 took the form of a letter designed to kill the recipient as soon as he opened it, but the plot was discovered in time.

The poisoner's art reached its highest achievements in Italy during the heyday of the Borgias. The claim that this family knew the secret of a subtle and deadly venom, beyond the reach of modern science, can hardly be substantiated, but poisoning was certainly a hazard among the nobility and clergy of sixteenth-century Rome.

... the heyday of the Borgias.

Jacob Burckhardt, the Swiss historian of the Renaissance, wrote of a white powder of an agreeable taste which could be mixed with the contents of any dish or goblet. This substance was probably white arsenic (an oxide of the metal), which was discovered by the Arab alchemist Jabir ibn Hayyan (or Geber) more than 1000 years ago. This material has no taste or smell, and a lethal dose does not take up much space. The Borgias did not need anything better. Rodrigo Borgia (the father of Lucrezia) was, according to a persistent legend, killed by his own poison, intended for a guest but switched by an incompetent or disloyal servant.

Arsenic remained a favourite instrument of homicide, even in England, where the law took a sterner view than it did in Mediterranean countries. Wriothesley's *Chronicle* for 1542 records the fate of one poisoner:

> This year, the XVII March, was boyled in Smithfield one Margaret Davis, a maiden which had poisoned three households that she dwelled in.

Until about 200 years ago, the chemists were well behind the criminals, for there was no reliable way of detecting arsenic in the body after death; indeed, it was not until 1845 that a satisfactory test was established.

Monarchs are now allowed to die of natural causes. But the poisoners can claim a distinctive place in the evolution of the modern world, and the pages of history will be less appealing if clever physicians find a humdrum explanation for every sudden death in a palace.

The Head Man

> I never satisfy myself until I can make a mechanical model of a
> thing. If I can make a mechanical model, I can understand it.

The confession of Lord Kelvin (then plain Sir William Thomson), spoken
in Baltimore nearly a century ago, is often quoted in support of the practical
man's resistance to mere theories. Kelvin did not go round with a bag of
nuts, bolts and Meccano spares; his models were made in the mind—or,
perhaps, in one of the famous green notebooks which he filled during train
journeys and dinner parties.

Scientists have been making models for nearly 400 years, and
doctors have been at it even longer. The best models are always based on
the latest fashionable technology. At the beginning of the nineteenth century, the eye was confidently described as a telescope, but later turned out to
be more like a camera. The brain was once a telephone exchange, but is now
a computer. Some of the most interesting models were made more than 200
years ago, when the new mechanics of Newton and Galileo were exciting
stuff. The bioengineers of those days saw springs in the muscles, levers in the
bones, bellows in the lungs, and shears in the teeth; their descendants today
use computers to make models, which are more sophisticated, but not
always more useful.

A once fashionable model, often derided but not yet dead, regarded
the brain as a jigsaw puzzle. This model is now popularly regarded as
disreputable because of its association with the cult of phrenology, but it
contains more than just a grain of truth.

Phrenology was inspired by the work of Franz Josef Gall, a German
physician who died about 150 years ago. He made four propositions, two of
which would not be disputed today. The brain, he said, is the seat of the
mind—or, at any rate, the means by which the mind communicates with the
external world. Different mental activities and attributes, he continued, are
associated with different parts of the brain. After that, he went wrong,
believing that parts of the brain grew in size according to the efficiency with
which they discharged their respective functions, and that this growth could
be detected by studying the shape of the skull.

Gall was a great anatomist, rightly respected for his careful dissec-
tions of the brain. As a scientist, he showed (long before Darwin) that the
brains of animals differed from those in men in degree, but not in basic
design. He was also a pioneer of criminology, believing that law-breakers
were the victims of the temperament imposed on them by their brains.

His ideas were to form the basis of a new materialist philosophy,
based on the observation of nature rather than on speculation, a possibility
which greatly attracted Herbert Spencer and influenced later philosophers.

Gall described the jigsaw puzzle of the brain in some detail,
identifying organs of amativeness, alimentiveness, veneration, language and

a score of others. His followers made phrenology into a superstitious cult, with its own temples, high priests and scriptures.

A copy of the first volume of the *Phrenological Magazine*, found a while ago in a second-hand bookshop, gives a diverting picture of the art as it was in 1880. A puzzled correspondent wrote:

> In a recent examination I found an unusually well-developed brain, the moral region splendidly exhibited, intellect very fine and the social organs good, with a well-balanced nature and vital temperament ... but was much surprised to find the man ... a stupid, commonplace character.... Could you explain this apparent anomaly?

The expert explained disdainfully that size is not everything, and finished off the awkward questioner with a barrage of meaningless jargon: 'tone, quality and texture of organisation are equally important.'

'much surprised to find the man ... a stupid, commonplace character.'

The same issue records an important judgment given in 'the Scotch Law Courts.' William Henderson, who died in 1832, left £5000 in trust for the advancement of the science of phrenology. In 1879, two of his nephews claimed the residue of the estate on the ground that phrenology was not a science. But the court decided against them, and this broad view of science was upheld on appeal. The Henderson Lectures, seeking to reconcile phrenology with anthropology and anatomy, continued into the 1920s.

Phrenology is dead, but today's neurologists still make maps of the brain, with the various functions, such as vision and hearing, quite sharply localised, thus confirming the basic soundness of Gall's ideas.

Livingstone the Doctor

Livingstone went to Africa as a missionary, but he soon became an explorer and finally a reformer, dedicated to the overthrow of the slave trade by the foundation of colonial settlements to encourage legitimate industry and commerce. His achievements (and failures) in these vocations have overshadowed his work as a doctor.

There was no General Medical Council in 1836, but Livingstone's training was adequate for the days before biochemistry, bacteriology or anaesthetics. He spent two winters attending lectures at the Andersonian University in Glasgow, and (after a year of theological study) another year at Charing Cross Hospital in London. He could not afford the fees for the final examinations and observed philosophically:

> ... it's no great matter—I shall be able to practice medicine among the Bechuana as well without as with the Licence of the Royal College of Surgeons.

The London Missionary Society thought that he should have a qualification, so he returned briefly to Glasgow (where the fees were lower) and obtained the Licence of the Faculty of Physicians and Surgeons.

Livingstone was not always an inspiring preacher, but he saw the practice of medicine as an important part of his ministry and soon gained a well deserved reputation as a physician. He was the first medical missionary in central Africa, and tackled his work in a spirit of humanity and scientific enquiry. He treated the witch-doctors with respect and courtesy, testing their remedies on himself and helping them with advice—though never in front of their patients.

He treated the witch-doctors with respect and courtesy.

Restless in the routine of the mission station, he was glad of opportunities for exploration, where his medical skill was decisive. Earlier explorers into Africa suffered heavy mortality (sometimes 100%) from malaria. Livingstone did not know the cause of the disease, though he made the significant observation that mosquitoes were abundant wherever malaria occurred.

His treatment was bold and correct—large and frequent doses of quinine. As a concession to the magical influences from which medicine was still emerging, the drug was mixed with ferocious purgatives. Livingstone's Pill (also known as the Zambesi Rouser) was very successful in keeping down the casualty rate. Livingstone was using the clinical thermometer and the wet pack in the treatment of fevers as early as 1852, when these techniques were still novelties in Britain. Sometimes, however, his enthusiasm for clinical science outstripped his knowledge. Faced with a smallpox outbreak in 1858, he wrote:

> I purpose to inoculate a cow ... in order to get the vaccine virus from it.

He was not sure what to do next, but the inoculation did not take and the experiment was abandoned. In 1849, having read of Simpson's work in Edinburgh, he wrote:

> Wish much I had some chloroform ... should have attempted to make some by a makeshift retort but fear the heat is too great here and it is very volatile.... Could we not procure some chloral and a retort?

Livingstone, like all the great explorers, was an eccentric, but his combination of curiosity and obstinacy would have made him a successful (if unconventional) physician anywhere. As it was, his medical work made a distinctive contribution to the opening up of Africa.

Electric Chair

Glasgow University benefits greatly from the generosity of alumni and their relicts. Books, paintings, manuscripts and gowns arrive in a steady stream to enrich the archives, galleries, libraries and ceremonial wardrobes. Some other resting place will have to be found for the gift recently noted in the annual report of the Court as

> ... the chair associated with the Old College in which members of the Anatomy Department are reported to have attempted to resuscitate a murderer's corpse by galvanisation.

'The chair ... in which members of the Anatomy Department are reported to have attempted to resuscitate a murderer's corpse.'

The Glasgow Medical School never supported resurrectionists of the calibre of Burke and Hare, but relied on the nocturnal enterprise of students in the anatomy class—often helped by their teachers. In 1814, Granville Sharp Pattison (a lecturer in anatomy), Andrew Russell (a lecturer in surgery), and two students stood trial for 'felonious abstraction of a body from the Ramshorn Churchyard.' They were acquitted—not without suspicion of trickery—and Pattison left hastily for the United States.

The criminal activities of the anatomists were effectively discouraged by these events, but the cause of learning received aid from an unexpected quarter. Matthew Clydesdale, a Lanarkshire miner, killed an old man with a pickaxe in August 1818. At the conclusion of his trial, the Lords of Justiciary

> ... decerned and adjudged that he shall be fed on bread and water only, till the day of execution, and that his body, after being so executed shall be delivered up by the Magistrates of Glasgow, of their officers, to Dr James Jeffrey, Professor of Anatomy in the University of Glasgow, there to be publicly dissected and anatom-ised.

... relied on the nocturnal enterprise of students in the anatomy class—often helped by their teachers.

On 4 November 1818, the murderer's body was taken, with an escort of troops and town officers, up the Saltmarket past the Cross, and into the Old College in the High Street. The Anatomy Hall was crowded, and the excitement increased when it became known that Professor Jeffrey, with the help of Andrew Ure (a teacher of chemistry) and other colleagues, was to experiment on the corpse with a new galvanic battery.

The scene which followed was so remarkable that the contemporary account would hardly have been believed had it not come from the incor-ruptible pen of Peter Mackenzie, the great social reformer and chronicler of nineteenth-century Glasgow.

The body, he related, was 'placed in a sitting posture in an easy armchair,' facing the audience. Two tubes attached to bellows and electrified by the battery, were placed in the nostrils. Clydesdale's chest heaved, his arms and legs moved and (Mackenzie assures us) he rose from the chair and stood upright.

Some of the students screamed, some fainted, and 'others of a sterner class clapped their hands as if in exultation at the triumph of the galvanic battery.' The medical faculty stood for a moment in amazement.

Then Professor Jeffrey took out his lancet and Matthew Clydesdale was executed a second time.

After that the judges consigned no more subjects to the dissecting room, though the demand remained brisk. Edinburgh used the services of Burke and Hare until the law intervened in 1828, but Glasgow proceeded more discreetly.

The Glasgow Medical School was of good repute and attracted a great number of students. Often the anatomy class exceeded 200, but the supply of subjects was adequate: 'the abundant communication with Ireland partly accounted for this,' as James Coutts cautiously observed in his history of the university, published in 1909.

Jeffrey took a prominent part in the campaign to legalise the practice of anatomy—a movement which culminated in the Anatomy Act of 1832. He served as a professor for almost 58 years (a record surpassed only by Patrick Cumin, who occupied the chair of Oriental Languages from 1761 to 1820), and was succeeded by Allen Thomson.

The new professor arranged for the faculty to buy Jeffrey's anatomical collection, which included numerous skulls and skeletons. No mention was made of the galvanic battery, but the easy armchair probably remained in the Anatomy Hall, for the recent gift to the university was made by Thomson's grand-daughter, Mrs G J Robertson, of Connel, Argyll.

Clydesdale will not be forgotten and the chair from which he rose, electrified, to meet his doom occupies a distinctive niche in the gruesome history of anatomy.

John Dee and the Armada

Agent 007 was busy on secret missions long before Ian Fleming created James Bond. The original master spy was Dr John Dee, an Elizabethan alchemist and astrologer, often dismissed as a charlatan, but now known to have accomplished some remarkable feats of military intelligence†.

Born in 1527 and educated in Cambridge, Dee taught Greek at Trinity College, and first became celebrated as a magician through the spectacular stage effects which he contrived in a production of a play by Aristophanes.

When Elizabeth I came to the throne in 1558, Dee (who had encouraged her during the previous reign by providing optimistic horoscopes) was commissioned to calculate an auspicious data for the coronation. The new Queen was not sure that she would be accepted by the people, but all went well and Dee's authority was confirmed.

... providing optimistic horoscopes ...

The Queen addressed him affectionately as 'My Ubiquitous Eyes,' a title which was reflected in the two circles of his emblem. The figure 7 had magical properties and was used particularly to denote familiarity with the occult world.

The combination exactly described Dee's technique. He was the Queen's eyes, watching for actions that might affect her interests, and his intelligence reports were invariably disguised as conversations with angels or other messengers from the spirit world.

† Richard Deacon 1968 *John Dee* (London: Muller).
Peter J French 1972 *John Dee: The World of an Elizabethan Magus* (London: Routledge and Kegan Paul).

Dee saw visions while gazing into a globe of smoky quartz or a mirror of polished cannel coal. The spirits sometimes appeared in the crystal but often materialised in the room. Since the person who invoked the spirits remained in a trance the messages could be recorded only by a team of two. Dee claimed to have written down the messages relayed by his assistant, Edward Kelley, who was employed as a seer (or scryer), but some of his judgments and predictions were more probably consciously prepared and wrapped in occult trappings to avoid premature disclosure, to add conviction, or to provide a cushion against possible error.

On 5 May 1583, the angels were asked:

> As concerning the vision wch yesterday was presented to the sight of E K as he sat at supper with me in my hall—I mean the appearing of the very great sea and many ships theron and the cutting off the hand of a woman by a tall, black man, what are we to imagine thereof?

The answer came from Uriel, the angel:

> The case did signifie the provision of forraine powers against the welfare of this land which shall shortly be put into practice, the other the death of the Queen of Scotts is not long unto it.

Dee's annotation in the margin was a drawing of an axe with the words: 'Q of S to be beheaded.' The fate of Mary Stuart might perhaps have been guessed by any competent student of politics, but the other prediction is more significant. It referred to the preparations, then just begun, for the building of the Spanish Armada. Walsingham, the official head of Elizabeth's intelligence service, heard nothing from his own agents until Thomas Rogers landed at Dartmouth in December 1585, with news of ships being assembled in Spanish ports. Dee's disclosure two years earlier may have been clairvoyance, but was more probably an indication of skilful and enterprising espionage.

Even more remarkable was Dee's work in foiling a Spanish plot which came near to destroying England's naval strength. In 1585, while in Prague, he heard from Francesco Pucci (another student of the occult) that a party of Frenchmen, in the pay of Philip of Spain, were to undertake a mission aimed at preventing the English from building any more ships. They were to arrive at the rendezvous 'before three yeares were ended and nine men beginne their Perambulations.'

Dee sent a messenger post-haste to London with the record of an angelic conversation. Madimi, a visitor from the spirit world warned: 'I see the risk of fire, great fire.' Dee asked: 'What is the significance of the three yeares and the Nine Men?' and was told: 'The Nine are Guardians of the RFD. They must acte against fire by the creatures of the Scorpion.'

The Queen's advisers soon deciphered the message. The Royal Forest of Dean was the main source of timber for the navy. The Nine were the committee of foresters about to start their triennial visitation. The enemy agents intended to set fire to the forest and to ruin the naval building programme.

The Verderers were alerted, the Spanish agents were caught with plans in their hands, and the danger was averted. Dee continued the campaign, using astrological predictions as psychological weapons against the promised Armada, and successfully predicted the downfall of Spanish ambition. He achieved much else during his long career, but the single-handed feat of counter-espionage that he wrought in Prague transcends even the successes of the later 007.

I Taught Wittgenstein

... I didn't teach him much, but the experience was interesting, and I learnt a little in the process.

Wittgenstein, one of the most influential scholars of the twentieth century, was appointed Professor of Philosophy at Cambridge in 1939. He found it difficult to teach, or even to think, during the war and sought menial employment as a porter at Guy's Hospital in London. He resisted all suggestions that he should take a more fitting job, but was eventually persuaded to transfer to a Medical Research Council unit in the same hospital, as a laboratory orderly. The unit had been established to study shock, but found itself under-employed when the predicted holocaust did not materialise. After a while the team moved to Newcastle, where accidents in shipyards and coal mines provided a modest supply of suitable patients.

Wittgenstein went with them, maintaining his frugal life style. Offers of accommodation from former students and fellow philosophers were declined and he settled for a hard bed, a wooden chair and a bare floor. His characteristic humility expressed itself in other ways: the librarian of King's College (now the University of Newcastle) was astonished when an assistant brought him a recommendation form in which two junior doctors certified that Ludwig Wittgenstein was a suitable person to enjoy the privilege of reading in the library.

He did, however, start to take an interest in the scientific work of the research unit; not unreasonably, for he had been a competent engineer and a good mathematician in Austria before he turned to philosophy.

... not a practical possibility.

I was just as astonished as the librarian when the professor brought Wittgenstein into the laboratory where I was working and asked if I would advise him on a little problem in electronics.

'I have found a paradox,' said the philosopher, 'and I cannot understand it.' He had been building an amplifier and stopped to consider how it worked. He saw that the input signal was applied to the grid of the valve and that the output appeared across an electrical resistance joined between the anode of the valve and the battery. The amount of amplification depended on the magnitude of the resistance. 'If I increase the resistance,' he said, 'I get more amplification?' I said that was quite right. 'Then, if I increase the resistance without limit, I ought to get unlimited amplification, even to infinity—which is ridiculous.'

I explained that what he said was theoretically correct, but not a practical possibility. Some of the battery voltage was used up in overcoming the resistance in the circuit, so as to keep the anode at a high enough voltage for the valve to work. So infinite amplification could be obtained only with an infinite battery.

He looked silently at the diagram that I had drawn, then looked at me. 'I see,' he said, beating his breast, 'I am a bloddy fool, a bloddy fool.'

Follies and Conceits

Divine Geometry

During the uneasy peace that followed the Peninsular Wars, Commander William Smyth, RN, was busy surveying the Mediterranean coasts of Europe and Africa. He was anxious when a Spanish ship came over the horizon, but found the visitor's intentions to be friendly. When the Spanish captain left, he presented his host with a handsome silver dish. Smyth was nonplussed, but restored the situation by handing over a set of bound volumes of his *Nautical Almanac*. The gift was in mint condition, for the *Almanac*, edited by the polymath Thomas Young (eminent in physics, medicine and Egyptology), was so full of mistakes that no British mariner ever relied on it. The Spaniards sailed away and were never seen again. Smyth, using French and Italian navigational tables, returned to home waters and died an admiral.

His skill in handling wrong ideas was inherited by his son, Charles Piazzi Smyth, who became Astronomer Royal for Scotland at the age of 26. He was an able astronomer but spent much of his time in studying the Great Pyramid of Giza.

The Pyramids of Egypt, built nearly 5000 years ago, have fascinated scholars since Herodotus wrote of his travels along the Nile. He thought that the Pyramids were tombs; other travellers described them as granaries, astronomical observatories, refuges from floods or dykes to keep the desert sands from spreading over the cultivated land beside the Nile.

The idea that the Great Pyramid was an elaborately coded message, recording past and future history, originated with John Wilson who, in the 1850s, claimed that the dimensions of the Pyramid were related to the distance of the Earth from the Sun, and to the chronology of the Bible. In 1860, John Taylor, a London publisher, proclaimed that the Pyramid had been designed by Noah and that its dimensions were based on the sacred cubit which had been used in the construction of the Ark, the Tabernacle and the Temple of Solomon. Smyth supported Taylor's ideas, although he thought Melchizidek, King of Jerusalem, to have been the architect. In 1865 he took an expedition to Giza and made a long series of measurements. He also photographed the interior passages for the first time, using a magnesium flash.

Smyth observed that the ratio of the length of each side of the Pyramid to its vertical height was almost exactly half of π. Since the Hebrews could not have known the relation between the circumference and the diameter of a circle, the Pyramid builders must have been divinely inspired. Smyth also claimed that the length of the Pyramid, divided by 365, yielded a measure of about 25 inches, which he identified as the Sacred Cubit. He calculated that 20 million of these cubits would equal the length of the Earth's axis and that the distance of the Earth from the Sun was a thousand million times the height of the Pyramid. Warming to his task, Smyth showed how other dimensions revealed the density of the Earth, the date of the Creation and the Flood, and the basis of a new scale of temperature. Furthermore, the British inch was related to the cubit and should not be replaced by the metric measures which were advocated even in those days.

... and made a long series of measurements.

Smyth's archaic units were much criticised, notably by Sir James Young Simpson, the pioneer anaesthetist (and amateur archaeologist) who reduced the Royal Society of Edinburgh to laughter when, at a meeting in 1868, he described Smyth's theories as

> ... a series of the strangest hallucinations, which many weak women believe, and a few womanly men, but no more.

Simpson went on to announce that the brim of his hat was equal to one twenty-millionth part of the Earth's axis. Smyth later described the scene indignantly:

> The pretended measurement was performed on the hat, in place of the sacred cubit of Moses as determined by Sir Isaac Newton; and performed with so much unction of manner and look as to be received with cheers by the large and learned audience.

Smyth's campaign was, however, encouraged by those (including Sir John Herschel, the astronomer) who took their tune from Mcquorn Rankine, the great engineer of mid-Victorian times:

> Some talk of millimetres and some of kilogrammes,
> And some of decilitres, to measure beer and drams;
> But I'm a British workman, too old to go to school:
> So by pounds I'll eat, and by quarts I'll drink,
> And I'll work by my three-foot rule.

Rankine was being facetious, but Herschel and his supporters were in earnest. Eventually their opposition prevailed and Britain had to wait for another century before embracing decimal coinage and making a start on metrication.

Smyth's technique—translating the dimensions of the Pyramid into chronology, past and future—has been practised and developed by many disciples. Mr Peter Lemesurier is the latest and certainly one of the most assiduous. His large and handsomely produced book† begins with the claim that the dimensions and materials of the Pyramid embody a coded message. Limestone means the way of the physical world; granite denotes eternity; ventilation shafts represent escape from mortality; a year is represented sometimes by an inch, sometimes by a fifth of an inch; 35·76 inches means reincarnation; 37·995 inches denotes death; and various other lengths or numbers relate to perfection, initiation, retribution, and so on.

Moving around and through the Pyramid with this elaborate code, Mr Lemesurier shows how the predictions cunningly concealed by the builders were verified in the invention of the printing press, the Diet of Worms, the American Tea Tax of 1767, the French Revolution, Marx's Communist Manifesto, and the war of 1914–18.

Other predictions for the future were less specific, including a new Messianic Age (for which preparation began in 1933); an irruption of the eternal into the temporal sphere in 1985; a messenger from eternity in 2039; and the start of a millenium in 2989. Earlier students of the Pyramid were sometimes more exciting in their predictions. Dramatic events, such as the end of the world, were confidently forecast for 1844, 1881 and 1953— but the cult still thrives despite these miss-hits.

Mr Lemesurier confirms his glimpse into the future by reference to the Bible and concludes with a summary of the Messianic Plan, 'whose purpose is to drive a motorway through the mountains of death and despair and to open up to man the fertile uplands of immortality.' As Alexander Woolcott remarked, after reading through a 12-page menu in a restaurant: 'Nothing there to object to.'

The Great Pyramid Decoded is not unique, but it is the most substantial recent offering of its kind. What inspires the production of these literary pyramids, involving much labour for no apparent purpose? The itch to write can hardly be invoked. True, Rossini declared that he could set a laundry list to music; but Isaac Newton, an early enthusiast for the mensuration of the Pyramid, cannot have been short of more weighty topics, nor indeed can Smyth.

Belief in the equality of man, and the consequent distrust of scholars, is perhaps a more important influence. Every shop steward knows how to run the country better than the Treasury economists. In an age when scientists are blamed for the real or imagined failures of technology, it is

† Peter Lemesurier 1977 *The Great Pyramid Decoded* (London: Compton Russell).

satisfying to believe that their incomprehensible theories can be circumvented by folk-wisdom or by powers which, belonging to the realm of the occult, need no explanation.

Respect for the occult is strengthened by the common belief that coincidences are unusual. In fact, they are very common, especially in the domain of numbers. The mysterious appearance of π in the dimensions of the Pyramid admits of a simple explanation, suggested a few years ago by T E Connolly, an electronics engineer. The Egyptians may have used a rolling drum for laying out long distances, such as the base of a pyramid. A drum with a diameter of one cubit would therefore produce a length of π cubits at each revolution. This technique would account for the observed ratio between the height and the length of the base of the Pyramid.

Anyone with a pencil and a pocket calculator can readily produce a string of uncanny coincidences. The number of days in the astronomer's year (365·242) divided by the square of π is 37·0—the body temperature on the centigrade scale. The length of Loch Ness is 22·75 miles or 40 040 yards, showing that there are ten monsters and that they have been there since the Creation. If we give a number to each letter, according to its order in the alphabet, and then add the values for Archbishop Ussher, Peter Lemesurier and Inverness-shire Monster, the result is 666, proving that the depths of the Loch are indeed occupied by a beast—perhaps even a Beast. With foreign travel so costly, the search for ancient wisdom might be continued nearer home.

Scientific Dowsing

The camera, the Asdic, and even the surreptitious rifle have not tempted the Loch Ness monster out of her cool retreat. Every year she makes a few brief appearances (after the song of the first cuckoo but before the arrival of the first tourist) to warn the hoteliers and the road patrols that both sides of the loch will once more be invaded by eager students of natural history.

Recently, a thoughtful monster-hunter concluded, from a study of earlier despatches, that frontal assault is unlikely to succeed. Accordingly, he concentrated his efforts on the enemy's lines of communication. Assuming, very reasonably, that the monster has come into the loch from somewhere else—and may dodge out again if the chase becomes uncomfortable—he used the ancient skill of the dowser to look for underground rivers. So far he has not succeeded.

The finding of a hidden stream is not a difficult task by the standards prevailing among those sensitive characters who respond to the radiations produced by water and by other materials even more desirable. The Abbé Mermet, who lived near Geneva, could spot an underground river, a spring, an oil well, or a plot of buried treasure from a great distance and (according to his admirers) was seldom wrong in his forecasts.

The monster ... would not have eluded his nimble pendulum for very long.

He used a pendulum made of a special alloy, which was held in the hand by a metal chain. The bob was hollow and could be filled with a sample of the material being sought, so that the radiations might more easily be recognised. In sensitive hands, the pendulum does not behave in the timid and impartial fashion suggested by the textbooks, but oscillates and gyrates in a complicated way which reveals (to the initiated) what lies underneath the ground. If the scene of the search is at an inconvenient distance, a map or photograph will serve. After some notable successes with swallowed buttons, the Abbé developed the use of the pendulum for locating truffles in Perigord, hares in coverts, predatory eagles, and lost cows. The Loch Ness monster, it must be presumed, would not have eluded his nimble pendulum for very long.

The Abbé Mermet's pendulum is one of the modifications by which the ancient craft of the water diviner has developed into the modern practice of radionics, which is modestly defined by one of its principal exponents as 'the science of detecting ultimate causes.' While the nuclear physicists mess about with their bubble chambers, and the biochemists make ever more complicated models of nucleic acid molecules, the dowsers labour in obscurity, guarding the basic stuff of knowledge and waiting for the recognition that they deserve.

There is, of course, more to dowsing than the pendulum and the twig. One of the rarer instruments is the fabulous fern seed which blooms only at the solstices. This has the property of disclosing treasure, for we have it on no less authority than that of Sir James Frazer:

> Whoever has it, or will ascend a mountain holding it in his hand on Midsummer Eve, will discover a vein of gold, or will see the treasures of the earth shining with a bluish flame†.

Another method is that the person fortunate enough to have found the seed should toss it into the air at midnight, when it will fall on the spot where golden treasure is buried.

Sometimes the fern blooms on Christmas Eve, when the golden reward is somewhat differently revealed. At that season, again on the stroke of midnight, if you can grasp the flower you will be able to force the Devil to bring you a bag of money.

The best divining rods are made from the mistletoe, which is, of course, the magical Golden Bough. It should be used as follows:

> The treasure-seeker places the rod on the ground after sundown, and when it rests directly over treasure the rod begins to move as if it were alive†.

Unfortunately, today's pirates and bandits hide their loot in Swiss banks, beyond the reach of the twig. But the dowsers try to keep their craft up to date.

Not all dowsers claim to have supernatural powers. During the past 100 years there have been many attempts to supply a scientific basis, usually related to some currently fashionable technology. The early enthusiasts invoked magnetism and electricity, but by the 1920s their successors were devising explanations based on radio waves. More recently, it has been

† Reproduced by permission of A P Watt Ltd, London.

claimed that the dowser is influenced by radioactivity in the materials which he tries to find.

... try to keep their craft up to date.

One system of dowsing is based on 'V-rays,' mysterious radiations unknown to orthodox science, but which are believed to be capable of travelling unlimited distances and of indicating the location of bank robbers or shipwrecked mariners with marvellous precision. A device embodying these unusual powers was warmly commended a few years ago in the official journal of H M Coastguard, published by the Board of Trade.

Unfortunately, the spectacular and unpredicted advances in technology during the twentieth century have gravely undermined the status of the supernatural. Broadcasting, photography and other innovations have given almost everyone powers greater than anything claimed or imagined by the magicians of old. Dowsing is a form of magic, seeking to explain real or supposed happenings by reference to powers possessed by the magician, but which are unavailable and incomprehensible to his client.

People who write ponderous books about dowsing often claim that their speculations have the support of this or that scientist, and must therefore be accepted as valid. This attitude of mind implies that science itself has a magical quality conferring immunity from critical assessment. But the practice of science is not at all mysterious; it involves no more than the application of intelligence and experience to a particular range of problems and activities. In many important respects, scientists are little different from other people. Sometimes they are gullible, and sometimes they find it hard to separate reason from emotion.

Is there an explanation for dowsing? Leaving aside the people who operate from a distance by swinging a pendulum over a map, there is no need to question the honesty or sincerity of the water diviners. It is true that they never submit themselves to properly controlled scientific assessment, but their successes can be explained without invoking trickery. A dowser will not find water in completely arid territory, but in many temperate lands there is an abundance of water not far below the surface, and therefore a correspondingly good chance of finding it at a reasonable depth, wherever the digging is done.

It is not disputed that the dowser's muscles cause the twig or rod to move. Often he is using geological clues based on local knowledge or general experience, and is unconsciously turning them into muscular movement. This process is well known among stage magicians, who can find a hidden object by holding the hand of the person who concealed it, or sometimes by merely watching the minute and unconscious muscular reactions of the bystanders.

Most water diviners are honest artisans providing a service which satisfies the people who buy it. There is nothing mysterious or supernatural about the way in which they do their work.

Scientific dowsing.

Stars in Daylight

It may be said that the fact makes a stronger impression on the boy through the medium of his sight, that he believes it the more confidently. I say that this ought not to be the case. If he does not believe the statements of his tutor—probably a clergyman of mature knowledge, recognised ability and blameless character—his suspicion is irrational, and manifests a want of the power of appreciating evidence.

Isaac Todhunter, the great Cambridge mathematician, was objecting to a foolish proposal to teach science with the help of experiments, and not as the philosophical exercise that had been good enough for Aristotle.

Since Todhunter's complaint of just over a century ago we have considerably improved the power of appreciating evidence. Today many people refuse to believe the mature and blameless clergyman, even when he does the experiment as well.

The problem goes back to ancient Greece by more than one route. Aristotle taught that stars could be seen in daylight by an observer at the bottom of a well or pit. The absurdity of this claim can be demonstrated by simple armchair reasoning (which, to be fair, was not available to Aristotle) and by direct experiment. But the legend survives.

Stars in daylight.

A celebrated book on the eye, first published in 1968, declares confidently that stars which are not normally seen during the day are visible from the bottom of a open mine shaft. An eminent scientist, writing in 1966, told of the ingenuity of Greek astronomers who worked in deep holes to avoid the inconvenience of waiting for nightfall. A textbook widely admired in universities advised: 'it is a well-known fact that an observer at the bottom of a deep pit shaft may see stars in the daytime.'

During the late 1960s, Dr Fergus Campbell, a distinguished physiologist, questioned the superstition in the pages of the *New Scientist* and asked if anyone had tried to verify Aristotle's assertion. This request did not evoke much response, but a subsequent reference in a Sunday newspaper produced the familiar result. Several readers wrote to say that stars could indeed be seen in daylight from the bottom of a pit shaft. Some claimed to have done the experiment themselves and others confidently quoted second-hand evidence.

In 1978, a writer in *Country Life* reported that workmen digging the foundations for Lancing College Chapel in Sussex, about 100 years ago, had seen stars at mid-day. After sceptical comments from a physicist, the story was affirmed even more confidently.

Why do we not see stars in daylight? Mainly because the sky is too bright. The eye's ability to detect a change in brightness depends on the background illumination; more exactly, the smallest difference in intensity that the eye can detect is proportional to the brightness of the background. On a dark night the sky delivers very little light, and the eye can detect the small increase produced by a star. In the daytime the background is much more intense and only a very bright star can produce a sufficient increment of light to obtain a response from the eye.

Stars can be seen in daylight with a telescope, but this effect depends on the lenses and not on the tube. If a star is to be seen, the eye must receive a certain quantity of light. The amount of light that can be collected by a lens (whether in the eye or in a telescope) depends on its diameter. The pupil of the eye (the exposed part of the lens) seldom has a diameter of more than half a centimetre. A very modest telescope will have an object lens five centimetres in diameter; that is, ten times the diameter of the pupil. The light-gathering power, depending on the area of the lens, will be 100 times greater.

The telescope has another lens, at the eyepiece end, which serves to magnify the image produced by the object lens. The image of a star is a mere point and is not affected by this process. But the light received from the patch of sky around the star is, through the magnification of the telescope eyepiece, spread over a larger area. Consequently, the background appears less bright when viewed through the telescope than when seen with the naked eye. The combined effect of the two lenses in the telescope is to make visible stars which would not be detected by the eye alone.

A pit shaft may be regarded as a long tube, but without lenses this does not help. Indeed, an observer whose eyes have become accustomed to the relative gloom at the bottom of a shaft will, on looking up, see the sky as exceptionally bright and will actually be less able to see stars than if he had remained above ground.

Apart from all that, a star could be seen only if it happened to be in the small patch of the sky visible from the bottom of the pit. There are only a few places in Britain at which bright stars are ever directly overhead: one in Sutherland, one in central Scotland, and two or three in England.

These considerations were patiently explained in 1916 by the Reverent W F A Ellison, an accomplished amateur astronomer of Waterford in Ireland. He showed that there was nothing to be gained (or seen) by digging holes at the locations mentioned, because the visibility of stars in daylight (except through a telescope) is a myth.

As though this was not enough, Mr Ellison actually made the crucial experiment, by descending a 900-foot pit shaft. He was struck not by the darkness of the sky, but by its brightness: 'If Sirius himself could have peeped down that shaft, his light would have been hopelessly lost in the glare.'

But fiction is stronger than truth and this preposterous legend, supported by eminent professors as well as by popular tradition, will not be extinguished by mere scientific explanation.

Drawing Magic from the Air

Most people would rather be cured by magic than by medicine; it's usually quicker and always more satisfying. The witch-doctors practise complicated magic—often with evident success—but the benefits are available even without travelling to Africa.

Most people would rather be cured by magic.

Ozone is a good example of homely magic. Everyone knows that seaside air is rich in this health-giving gas. In recent years, ozone generators have been on sale for use in homes, public places, and even sewage works. Ozone, it seems, destroys smells and germs; if these are not to be found, it freshens the air and generally invigorates bystanders.

Scientists and doctors are, as usual, somewhat unfriendly to these new ideas. They point out that there is no ozone at the seaside; indeed, ozone is not normally found in the atmosphere in significant amounts near the ground.

But what about the smell? It's not ozone, say the killjoys, just rotten fish and decaying seaweed. That may be, say the ozone-fanciers, but surely it would do good if there was enough of it about? Unfortunately, ozone is really a poisonous gas, more dangerous to life than chlorine, carbon monoxide or prussic acid. It's one of the principal irritants in the Los Angeles smog.

In animal experiments, susceptibility to pneumonia and other infections was markedly increased by exposure to quite low levels of ozone, sometimes no greater than would be produced by commercially available ozonisers. At very high concentrations, ozone does destroy germs—but at these concentrations it is likely to destroy people as well.

Less powerful magic, also warmly commended by many, is conveyed by negative ions. In this context, negative ions are formed when electrons attach themselves to clusters of water molecules. Electrons are always present, through the action of cosmic rays and of radioactive materials in the air and the Earth.

The air contains positive ions too, since a molecule of oxygen or nitrogen which has lost an electron (and therefore has become positively charged) may attach itself to a cluster of water molecules. Town air usually contains a few hundred ions per cubic centimetre—with the positives slightly outnumbering the negatives, which are repelled by the negative charge usually present on the Earth's surface.

Various devices are sold with the claim that they increase the concentration of negative ions in the air. The ionisation is produced by a radioactive source or an ultraviolet lamp; a fan and an air filter are sometimes added and ozone is occasionally claimed as a bonus.

These simple devices are reputed to destroy smells, germs, and even vermin. Relief of various afflictions (including asthma, hay fever and high blood pressure) is also claimed. Other benefits are said to include the elimination of mould in breweries and cheese factories, the control of distemper in animals, and the relief of pain after burns.

Negative ions are generally supposed to be beneficial, producing 'comfort, optimism, exhilaration, good temper, and friendly attitude,' while positive ions lead to 'fatigue, headaches, dizziness, nausea, and faintness.' But the reports are conflicting. One research worker found that cucumbers grew to surprising lengths under positive ion treatment, and two American physicians reported that some of their asthmatic patients wheezed when they breathed negative ions, but not when supplied with ordinary air.

The United States Food and Drugs Administration told an English engineer in 1966: 'Negative ion generators are ineffective in producing any therapeutic or health benefits.'

On the other hand, there are many satisfied customers quite convinced that ionised air has relieved their afflictions or banished bad smells. There is nothing odd about these differences of opinion or experience. The essence of magic is that it helps some people but not others. A cure that works for everyone is not magical, but merely scientific.

Meanwhile, the ions keep rolling. Cars become stuffy—so install negative ion generators to restore the weary drivers. Smokers cough and splutter—try ions to help their breathing. Ions are certainly more expensive than copper bangles or dangling chains behind the car—and they don't do much harm. But keep off the strong magic; whatever ails you, ozone won't cure it.

Diminishing Returns

Perpetual motion once held a respectable place among the follies of science, along with squaring the circle, trisecting an angle, the philosopher's stone and the universal solvent. The transmutation of metals has been achieved by nuclear engineers, but not on the profitable scale envisaged by the alchemists. The universal solvent is water. Palingenesy, the re-creation of plants or animals from their ashes, has its modern counterpart in the deep-freeze mortuaries to which the affluent may consign their remains in the hope of eventual revival. But perpetual motion is no longer very interesting. Today's vulgar errors—ESP, UFOS, spoon-bending and talking to plants—are not intellectually stimulating, since they demand merely the suspension of reason.

Today's vulgar errors ... talking to plants ...

Circle squaring was an ancient craft (Archimedes looked at the problem), but perpetual motion did not excite attention until 500 years ago, when the Renaissance engineers began thinking about machines and mechanisms. The miller's water wheel turned unceasingly as long as the river flowed, but how could the corn be ground during a dry season? A tankful of water would move the wheel for a short time. Archimedes had shown, in the third century BC, how water could be raised by a screw turning in a tube. So, if the wheel drove the screw, the water could be lifted and used again. More fancifully (as Zimara, a sixteenth-century Italian physician suggested), a windmill might be used to work bellows which would produce enough draught to keep the sails turning.

The early perpetual motion machines never left the drawing board, and not all of them pretended to do anything useful. Apart from the devices driven by water and wind, there were many schemes for amusement only. One design, which had many variations, envisaged a vertical wheel with metal balls running in grooves, arranged so that one side of the wheel was always heavier than the other, so ensuring a continual rotation.

This kind of perpetual motion machine appeared plausible, because the Sun and Moon kept going with nothing obvious driving them. But inventors now look for profit. Since there is not much money in aimless movement, many efforts have been made to produce purposeful machines. Only a short time ago a Scottish newspaper reported the exploits of a company director who, by mounting a fan on the roof of his car, generated enough electricity to run his home and to halve his petrol consumption. A few years ago the same newspaper told of an inventor who put the fan on the bridge of his boat, with spectacular improvements in speed and fuel economy. During the summer of 1976, a company in California unveiled a black box which, without external power supplies, converted water into hydrogen, thereby delivering fuel for nothing. Shares did well until dealings were suspended. However, perpetual motion is no longer a growth industry. Its heyday occupied the second half of the nineteenth century, when nearly 600 patents were granted in England. These, like all the other proposals that have been or will be made, can be classified either as fallacies or as frauds.

The fallacies are schemes which seem to circumvent the basic law of mechanics (nothing for nothing), but only because the operation of the law is not sufficiently appreciated by the inventor. Everlasting clocks provide several examples of ingenious fallacies. A carbon and a zinc rod, buried a foot or so apart in damp soil, will generate enough electrical energy to drive a clock. One model ran for 40 years in the Leicester Museum, and a few hundred were built before production ceased in 1914. More than 200 years ago, James Cox of London built a serviceable clock driven by changes in atmospheric pressure. It is now in the Victoria and Albert Museum, though it is no longer working. Eighteenth-century inventors made clocks driven by the expansion and contraction of metal rods in response to variations in temperature; today there are clocks driven by solar cells, which convert sunlight into electrical energy.

These ever-moving mills, wheels and clocks all fail because of two errors—one simple and one profound. The simple error resides in the inescapable fact that cog wheels, axles and ratchets all lose a few million atoms every time they move and must eventually wear away. This takes a long time, but not as long as the eternity which the perpetual motivators claim.

The death knell of perpetual motion was formally delivered in 1852 by William Thomson (afterwards Lord Kelvin) the 27-year-old Professor of Natural Philosophy at Glasgow University. In a monumental paper read to the Royal Society of Edinburgh he reached two main conclusions:

> 1. There is at present in the material world a universal tendency to the dissipation of mechanical energy.
> 2. Any restoration of mechanical energy, without more than an equivalent of dissipation, is impossible . . . .

In reaching these conclusions Thomson explained, in effect, that mechanical or electrical energy could be degraded (to heat) without loss, but that the reverse process could not be achieved without an additional supply of energy. When high-grade energy (mechanical, electrical, chemical or nuclear) has been converted to heat (whether by friction or in any other way), it cannot be restored except at an unfavourable rate of exchange. Since all machines turn high-grade energy into heat (which is the lowest grade), the total amount of energy in the universe, even though remaining constant in quantity, must inevitably deteriorate in quality. Not only is perpetual motion impossible, but the universe is moving inexorably towards the ultimate perpetual stillness, with the whole of creation at a uniform temperature and no available energy anywhere.

We can, of course, make local and temporary exceptions to this process by cashing in on oil to make electricity and more sophisticated forms of energy, but the ultimate fate is inescapable. Indeed, everything that we do hastens the end. The wind blows freely, but every windmill removes a little of its energy (eventually to be dissipated as heat) and slows the Earth down. The zinc and carbon rods of the primitive electric clock wear out and can only be replaced by the consumption (in a chemical factory) of more energy than they delivered while intact.

Mr Ord-Hume, in an interesting book† which is a by-product of a television programme, describes a great many perpetual motion schemes and explains their defects. He does not accept the scientific view of perpetual motion, asserting instead that the futility of the concept can never be proved but can only be established in particular cases by detailed examination of drawings or models. Here he does less than justice to the early scientists, including Leonardo da Vinci and William Gilbert (physician to Queen Elizabeth I of England), who understood the folly of the search for perpetual motion even before the theoretical reasons were produced. But Ord-Hume's book is a valuable record both of well meant effort and of deliberate fraud. His stories of the bogus power schemes (all based on hidden power sources) of the nineteenth and even the twentieth centuries are well documented studies in gullibility.

† Arthur W J G Ord-Hume 1977 *Perpetual Motion: The History of an Obsession* (London: Allen and Unwin).

The Cloudy Crystal

The Veterans of Future Wars were a group of young Americans who came together after 1945, explaining that they would like to have the benefits of military adventure (such as housing loans and free university education) while they were still alive and healthy.

Amateur historians have recently discovered that it is more fun to write history in advance than to wait for it to happen. Already there is a Commission on the Year 2000, an Institute for the Future, a World Future Society, and many journals devoted to the guessing game.

They thrive on the realisation that no one objects if the fortune teller is wrong—as he usually is. Even the experts are often adrift in their forecasts. Lord Kelvin wrote in 1896 (to Colonel Baden-Powell): 'I have not the smallest molecule of faith in aerial navigation other than by balloon rig.' Simon Newcomb, the eminent American astronomer, observed a few years later: 'No possible combination of known substances, known forms of machinery and known forms of force can be united in a practical machinery which man shall fly long distances.' It seems like only the other day (although it was actually in 1956) that the newly appointed Astronomer Royal described space travel as 'bilge, utter bilge!'

'I have not the smallest molecule of faith in aerial navigation . . .'

Forecasters are often too timid. Around the turn of the century, the popular prediction told of food in the form of pills, and that escalators would be the transport system of the future. Sometimes they are too optimistic, but most often they are simply wrong.

Occasionally the predictions survive long enough to be checked in detail. In 1936 John Langdon-Davies wrote *A Short History of the Future*†.

† John Langdon-Davies 1936 *A Short History of the Future* (London: Routledge and Kegan Paul).

Democracy, he thought, would be dead by 1950. By 1960, work would be limited to three hours a day; the school leaving age would be 21; power would be so cheap that electricity would be virtually free; and food and clothing would cost 'as little as air.' By 1975 the family would have disappeared and the home would be 'reduced to a dormitory with good modern plumbing.'

In 1962 a battery of American experts wrote a large book describing what the world would look like in 1975. How do the predictions compare with the true state of things? Some of the forecasts were reasonable enough, although the scientists who foresaw the impact of electronics on the collection of income tax did not have a vision of the Inland Revenue computer at East Kilbride near Glasgow, which produces sackloads of rubbish every day, to the great annoyance of the victims and their accountants.

Other predictions were less successful. By the 1970s, it was confidently asserted, automatic translation of practically all foreign publications and books would be routine. This hope drove a well filled bandwaggon during the 1950s, but by 1966 the idea was abandoned. Language, it seems, is too complicated for a machine to understand; the instinctive way in which we choose among the many different meanings of a word or phrase cannot yet be written on magnetic tape.

Transport was expected to enjoy great improvements, with 150 supersonic planes in the air and 100-knot hydrofoils plying the oceans. The nuclear ship *Savannah* was to be the first of a fleet. Actually she turned out to be the last; shipbuilders were anxious to see the development of nuclear power, but ship-owners knew that the prospect was hopeless in financial terms. Pocket television receivers arrived only in 1978, but the 50-inch screen, also expected by 1975, has not yet moved from the drawing board to the drawing room.

What use is the future? On the whole, we don't take much notice of it. Warnings of environmental pollution or tobacco poisoning have turned out to be justified, but have not greatly influenced personal and public behaviour. One of the world's most eminent scientific societies has recently begun distributing its proceedings in plastic bags (which cannot be disposed of without polluting the environment) instead of old-fashioned biodegradable envelopes.

Predictions about the future will nearly always be wrong, but they may encourage us to think about the present. Mostly they don't because the forecasters are not sufficiently imaginative. Another Swift, another Butler or another Wells might stir the social and technological conscience but, unfortunately, today's science fiction (or, for that matter, today's science nonfiction) is not in the same league.

Fact and Fiction

The sale of alcoholic beverages was (by vote of the citizens) banned in Kirkintilloch in Scotland from 1928 to 1969. Since then the townsfolk have approached in reality the reputation which they already enjoyed in fiction.

During the years between the two world wars, the Provost and Town Clerk received letters from the ends of the Earth, seeking information—or even samples—required for the further study of Duggan's Dew of Kirkintilloch. This champion among whiskies was the invention of Guy Gilpatric, who wrote stirring yarns describing the adventures of Mister Glencannon and other bold seafarers.

Duggan's Dew was not only their favourite solace and stimulant, but had almost miraculous powers in soothing enemies, lubricating bureaucratic obstructions, or changing tense confrontations into amiable drinking parties—powers otherwise found only in the cans of spinach so effectively used by Popeye.

... samples ... required for the further study of Duggan's Dew of Kirkintilloch.

Readers of Gilpatric's stories found themselves with a powerful thirst after a few pages, and made haste to lay in a stock of the noble spirit. When Indian merchants, Australian bartenders, and African storekeepers protested ignorance, many of the seekers contented themselves with lesser varieties; but the more determined sought first-hand evidence, unwilling to believe that liquor so convincingly described could be a mere literary figment. Always the Provost had to explain in embarrassed words that Duggan's Dew of Kirkintilloch did not exist, and that even if it did, there would be none in Kirkintilloch.

Whether Gilpatric chose the name at random or whether he was moved by a spirit of innocent mischief is an open question, but there are still many corners of the Commonwealth in which Kirkintilloch evokes the same response as Glenlivet, Laphroaig, or Talisker.

Fiction is more often more compelling than truth. Sir Walter Scott took most of his principal characters from life, but transformed them into altogether nobler or wickeder people, and in doing so made them more believable.

But for Scott, Meg Merrilies would still be an obscure gipsy and Rob Roy a Highland ruffian awaiting the touch of life from a television producer. Lucy Ashton (who achieved musical fame as Lucia di Lammermoor) was somewhat too convincing.

For *The Bride of Lammermoor*, Scott used the legend of Janet Dalrymple (daughter of the first Viscount Stair), who is said to have stabbed her bridegroom on their wedding night and died insane a few days later. The facts are otherwise, but Scott's version was taken up by Macaulay, who turned it into respectable history.

Scott did not try to conceal his sources, but freely acknowledged his borrowings, even when they took the form of self-portraits—usually flattering, as in Frank Osbaldistone or Colonel Mannering, but occasionally in the satirical vein represented by Jonathan Oldbuck the antiquary.

When Scott wrote about the life of a lawyer or a country gentleman, he did at least have some experience to guide him. Trollope's Barchester novels, with their compelling picture of life in a cathedral town, were composed without any inside knowledge.

Trollope's work as a Post Office surveyor took him on one occasion to Salisbury, where the first of the novels was conceived during an evening stroll. The plot of *The Warden* owed much to newspaper reports of a litigation promoted by the Reverend Robert Whiston, headmaster of the Cathedral Grammar School in Rochester, who chased the Dean and Chapter through the Bishops' Court and the corridors of Chancery.

But the clerical characters, universally applauded for their authenticity, were entirely imaginary. Trollope had never met an archdeacon, nor lived in a cathedral town. His dignitaries were evolved by instinct—or, as he was fond of saying, by an effort of moral consciousness.

Authentic or imaginary, fictional characters do not usually outlast their creators. Sherlock Holmes enthusiasts, who have read their way through the canon with unquenched appetite, must simply start again; but television adds a new dimension to the production of fable and folklore.

The original Dr Cameron of Tannochbrae, who appears in A J Cronin's autobiography, had a small beard. Janet, the housekeeper was

> . . . a thin, elderly woman, dressed entirely in black . . . Her hair was tightly drawn, her person spotless, and in her bleak face was stamped authority, mingled with a certain grudging humanity†.

† From A J Cronin 1951 *Adventures in Two Worlds* (London: Victor Gollancz).

Dr Finlay stayed in Tannochbrae for little more than a year. Then Dr Cameron succumbed to pneumonia and

> ... there appeared at the funeral from a remote northern town, two nephews and their wives. They descended in apparent grief, in reality like wolves upon the fold. Nothing was sacred to them in their avaricious possessiveness, from the dead man's bankbook to the clothing in his bedroom cupboard.... within a month the practice had been sold†.

But the magic touch of radio and television gives charm to Janet and immortality to Dr Cameron.

Science and the Supernatural

The King of Syria was worried about leakage of information to his enemies. 'Will ye not show me,' he asked his staff, 'Which of us is for the King of Israel?' 'None, my lord, O King,' was the answer, 'but Elisha, the prophet that is in Israel, telleth the King of Israel the words that thou speakest in thy bedchamber.' This exchange, which we read in the second Book of Kings (chapter 6), provides an early claim for the existence of a facility now known as telepathy or extra-sensory perception.

'... Elisha the prophet ... telleth the King of Israel the words that thou speakest in thy bedchamber.'

The idea that psychic phenomena should be studied using the methods of scientific enquiry became popular about a hundred years ago and led to the formation in 1882 of the Society for Psychical Research. During the 1870s, Lord Rayleigh, the physicist, and A J Balfour, afterwards Prime Minister, were members of a group which investigated the supernatural, without much success, in Cambridge. Among the other scientists who took an interest in the Society's affairs were Sir William Crookes and Sir Oliver Lodge. Three of the early members, Gurney, Myers and Sidgwick, kept in touch with their successors by lengthy messages, mostly in the form of literary puzzles transmitted with the help of mediums.

Telepathy, the subject of many recent enquiries, owes its name to

Myers, who was a classical scholar at Cambridge, and afterwards an inspector of schools. It has been defined by Sir Alister Hardy, the zoologist, as 'the communication of one mind with another by means other than by the ordinary senses.' Experiments in telepathy, using playing cards, were made with great success in 1881 and 1882 by Sir William Barrett, a physicist, aided by the daughters of the Reverend A M Creery. The girls may have been too clever, for they were afterwards caught cheating in a similar experiment. Sir Oliver Lodge proposed systematic methods of testing telepathy by card guessing in 1885, and explained how the significance of the results might be assessed mathematically.

The first experiments of this kind were made in 1917 by Professor John E Coover, an American psychologist, and 200 of his students. Coover took cards one at a time from a pack and handed them to a student sitting beside him. Sometimes the student looked at the card and concentrated on it, while an observer in a different room recorded his guess. On other occasions, decided by the throw of a dice, the first student took the card without looking at it while the second student made his guess. About 5000 trials were made in each of the two series and Coover reported that neither of them showed success significantly above the level expected to occur by chance. Other people have rearranged his results and claimed that they do show slight evidence of telepathy.

Another experiment, done by Miss Ina Jephson in 1924, used 240 subjects in postal communication with the organiser. Each of them was asked to draw a card from a pack, guess its value before looking at it, and write down the guess along with the correct answer. This test was repeated until five cards had been drawn, and the same process was gone through on five different days. The results analysed by Miss Jephson showed correct guesses much in excess of the chance expectation.

As described in some widely read books on telepathy, this was a successful experiment. What is not always explained is that most of the 240 subjects did the tests in their own homes with no supervision. The possibilities for error are therefore obvious. In a more foolproof experiment made a few years later, the subjects received by post a series of playing cards in opaque envelopes. Having guessed the contents, they returned the envelopes unopened. Careful precautions were taken to prevent any form of cheating. The results were completely negative.

Between 1927 and 1929, the BBC cooperated with the Society for Psychical Research in attempts to discover telepathic powers among listeners. These tests were inconclusive, as were similar experiments made in the United States. Playing cards are not very good material for this kind of investigation. When 24 000 British listeners responded to an invitation to guess a card, about 1000 chose the ace of spades and nearly as many chose the nine of diamonds, known to card players as 'the curse of Scotland.'

Much of the later interest in telepathic and such like effects sprang from the work of J B Rhine, who was active in these matters for more than 30 years at Duke University in North Carolina. He used Zener cards in packs of 25, each card bearing one of five simple designs: circle, square, star, cross or wavy lines. The experimenter removed cards one by one from a shuffled pack and the subject recorded his guesses, either in the same room or at a distance. In the early years of his investigations, Rhine came across many sensitive subjects who, he claimed, could demonstrate telepathic powers, but few were found after 1940, and most of the original subjects lost their gift after a while.

In England, Dr S G Soal, a London mathematician, made a long study of Rhine's experiments and carried out several trials with British subjects using Zener cards or others bearing coloured pictures of five different animals. Soal was at first sceptical of Rhine's claims and accordingly designed his own experiments with many precautions against error or deception. His most successful collaborator was Mrs Gloria Stewart, who made an average score of about 25% in 37 000 guesses, compared with a chance expectation of 20%. Mrs Stewart's powers declined in 1949 and did not return. During the 1950s, Dr Soal made a series of trials with two Welsh schoolboys. Encouraged by substantial gifts of money for correct guessing, the boys achieved extraordinary success, but were caught cheating in some of the tests.

The experimenters offer five arguments in support of their claims. Firstly, they explain that modern scientists are making their enquiries ever more scientific; for example, by using electronic devices such as random-number generators.

Secondly, they produce lists of eminent scientists and philosophers who have studied the supernatural and have held office in the Society for Psychical Research. Thirdly, they assert that modern physics no longer depends on common sense, but dabbles in time reversal, negative matter and other irrational concepts; so ESP is as good as science. Fourthly, it is claimed that telepathic communication must soon become commonplace because (as Arthur Koestler put in in 1972) 'science fiction has proved to be an astonishingly reliable prophet.'

These arguments are flimsy. Science is not made by illustrious patrons or technical trimmings. The test which distinguishes between science and non-science is quite simple. Any observation, however interesting or curious, remains an anecdote until it can be repeated by someone else. That is why respectable scientific journals regularly publish interludes of rubbish; of course, no one knows until a few years later which is the rubbish and which the genuine article, but time inevitably brings the answer.

When parapsychologists protest that modern physics is incomprehensible to them, so why should their efforts be judged by physicists, they are preaching the gospel that Ivan Illich offers in place of medicine— and that extremists offer in place of rational politics and economics.

Mostly, however, the parapsychologists plead, as their fifth argument, that they deserve to be taken on trust. Although their experiments cannot be repeated, nor demonstrated to order, they should still be treated as scientific knowledge, because the people who do the work are vouched for by genuine scientists. As Pliny observed, the cobbler should stick to his last; scientific ability is no guarantee of good judgment in gardening, politics or theology.

Scientific ability is no guarantee of good judgment in gardening.

The validity of ESP could, of course, be established in a very simple way. If some practitioner of the art would publicly and accurately forecast the outcome of elections, stock market changes or sporting events, all arguments would cease. No performance of this kind has ever been given and current research does not envisage anything so straightforward. The enthusiasts record large numbers of guesses about trivial issues, such as the designs on cards, and claim that any departure from the so-called laws of chance is proof of the supernatural.

These are feeble weapons of persuasion, which the great wizards of the past have scorned. Nostradamus and the rest of them made thumping big prophecies which could be proved true or false without slide rules and books of tables. Their modern counterparts will need to do something of the same kind if extra-sensory perception is not to moulder along with phrenology, palmistry and astrology in the dustbin of science.

Criticisms of experiments made to verify telepathy fall into two groups: the general and the particular. Objections to particular investigations take several forms. Rhine's early tests were done without proper safeguards against error. The guesser sat with him at a small table watching the drawing of the cards and the checking, sometimes after every five guesses. Small marks on the backs of the cards and a variety of other clues might have been used, consciously or unconsciously. Rhine acknowledged these weaknesses and explained in 1936: 'we work here on the principle that you must first catch your butterfly before you can pin it down.'

Naturally enough, only the more spectacular results were published, a process which incurred the wrath of H L Mencken, who complained: 'Professor Rhine segregates all those persons who, in guessing the cards, enjoy noteworthy runs of luck, and then adduces those noteworthy runs of luck as proof that they must possess mysterious powers.'

Professor Mark Hansel, a British psychologist, believes that some of the results obtained in British experiments are consistent with trickery. He claims that the remarkable powers exhibited by Dr Soal's subject, Mrs Stewart, disappeared in 1949 after a possible routine for dishonestly producing high scores had been described. He suggests that the performance of the Welsh schoolboys who took part in Soal's experiments could easily have been obtained with the help of a supersonic whistle, inaudible to adults, when they were not using cruder deceptions.

The general objections to experiments purporting to prove telepathy are numerous. The most obvious is that extra-sensory perception can seldom be demonstrated unless the people in charge of the experiment believe in it; the presence of sceptics is said to upset the delicate forces controlling the subject's ability to see what is hidden.

A simpler explanation is also possible. In a test made at Stanford University, 1000 card guesses were noted by an investigator convinced of the reality of extra-sensory perception. Correct answers totalling 229 were recorded—significantly more than the chance expectation of 200. Unknown to the experimenter, a tape recording was made during the proceedings. This showed that the percipient scored only 183 correct guesses, the other 46 being added erroneously by the experimenter. When the experiment was repeated without concealing the tape recorder, only two mistakes occurred.

Another criticism is that the parapsychologists sometimes do not decide what experiment they are doing until after they have collected a mass of observations and had a look at them. If a set of card-guessing trials shows a high score, that is evidence for telepathy without further argument. If a low score results, the subject may be demonstrating negative extra-sensory perception. If the score is just about what might be expected, a search is made by comparing each guess, not with the corresponding card, but with the previous one or the next one, or others even further removed.

An objection of a more fundamental kind may be summarised in the words used by William of Occam more than 600 years ago: 'it is vain to do with more what can be done with fewer.' This maxim (in a slightly different form) came to be known as Occam's Razor, from its usefulness in removing the superfluities of medieval philosophy. Its application to the problem of extra-sensory perception was anticipated by Tom Paine, who asked: 'Is it more probable that nature should go out of her course, or that a man should tell a lie?'

These and other arguments of a similar kind were presented by Dr George R Price of the University of Minnesota in a brisk attack on parapsychology made in 1955. 'If, then, parapsychology and modern science are incompatible,' he wrote, 'Why not reject parapsychology? We know that the alternative hypothesis, that some men lie or deceive themselves, fits quite well within the framework of science.' Price went on to explain how he could duplicate some of the famous telepathic experiments of the past 30 years, using confederates in such a way that the fraud could not be detected.

Objections of a more fundamental nature were made in 1957 by Mr G Spencer Brown, an Oxford philosopher. Parapsychologists always express the significance of their card-guessing results by quoting the odds against the observed scores occurring by chance. In the common test with 25 cards bearing one of five symbols, the most likely result is five correct answers. The probability (calculated mathematically) of scoring more than 10 out of 25 is about 150 to one. In other words, a score of this size may be expected to occur about once in every 150 trials during a really long series of tests.

Unexpected results no more probable than this sometimes turn up in scientific experiments, but they are easily disposed of. Anyone who reads something in a scientific journal which he finds hard to believe does not need to invoke supernatural effects to find an explanation. He can do the experiment for himself, using the same method as the original investigator. If he finds the same result, and if other scientists do likewise, the matter is proved. Its explanation may need new theories, but the experimental results need not be in doubt.

In parapsychology, this simple test cannot be done. The abnormal results in card-guessing are produced only by certain people at certain times, which cannot be arranged in advance. The curiously high scores cannot be reproduced at will, even by the original experimenters. Subjects who have shown the more remarkable successes have invariably lost their powers over a period of years.

This effect is of crucial importance, according to Spencer Brown. He points out that any succession of events occurring at random contains sequences which differ widely from what the laws of chance would suggest. He demonstrates the truth of this assertion by reference to the tables of random numbers prepared by statisticians and used in the design of scientific experiments. Even in this unpromising material he found numerous repetitions, omissions, and other patterns which are, at first sight, anything but random.

Some of these oddities were noticed by the compilers of the tables and were suppressed before publication, but others still remain. If the machines which make the random numbers were allowed to continue for a long enough time, the anomalies would disappear or give way to new ones. Psychical researchers have, says Spencer Brown, spent the past 70 years in trying to demonstrate something which does not exist.